Semiconductor Ecosystem Strategy

: Business Models, Market Forces, and Strategic Frameworks

The first book to introduce Semiconductor Business Case Studies of Apple, Google, NVIDIA, and Samsung

Semiconductor Ecosystem Strategy

Published and Printed in the Republic of Korea by Yonsei University Press.

50 Yonsei-ro, Seodemun-gu, Seoul, 03722, Republic of Korea
Tel: (+82-2) 2123-3380~2 **Fax:** (+82-2) 2123-8673 **E-mail:** ysup@yonsei.ac.kr
http://press.yonsei.ac.kr

ISBN 978-89-6850-910-0 (93560)

₩ 25,000

SEMICONDUCTOR ECOSYSTEM STRATEGY

A Groundbreaking Insight into
Semiconductor Business Models of
Apple, Google, NVIDIA, and Samsung

Seung Hwan Jeon

YONSEI UNIVERSITY PRESS

For Everyone in the Semiconductor Ecosystem

To the Lord, the Blessed Virgin Mary, and my

family and friends, I offer my deepest gratitude.

Special thanks to Samsung and Yonsei University

for providing this invaluable opportunity.

In honor of my father watching from heaven,

as well as the esteemed late chairmen.

Contents

Chapter 3.

Semiconductor Business Models and Strategies · 49

Chapter 4.

Conclusion · 145

References · 155

Introduction

Purpose and Key Insights

One of the pivotal contributors to the advancement of human civilization lies in the development of **semiconductors.** These tiny yet powerful devices not only sustain modern technological progress but also exert a profound influence on the individuals devoted to their creation, refinement, and commercialization. Some people devote the entirety of their careers to semiconductor technology and business operations, while others harbor ambitions to pioneer new semiconductor architectures or launch successful enterprises through advanced R&D. Still others perceive semiconductors primarily as an appealing—if at times intricate—area for investment or intellectual pursuit.

Although the recent ascendancy of **AI semiconductors** has captured significant attention, it nonetheless represents only one branch of the broader semiconductor domain. As technology evolves, competition grows increasingly formidable, prompting semiconductor organizations to operate **24/7** in pursuit of revenue and profits. These firms grapple with a convergence of formidable rivals, arduous pricing discussions with customers, and fluid market conditions. In this book, I seek to offer a window onto the narratives, experiences, ideas, and practices of the many players in this far-reaching semiconductor ecosystem.

Building upon extensive industry experience, coupled with a deep awareness of emerging trends, I have observed that while the **semiconductor sector** is often construed as primarily technology-centric, **examining it from a business vantage reveals patterns, principles and models integral to sustainable success.** Indeed, the industry comprises unique business models and foundational principles that demand carefully crafted strategies. In times marked by the rise of new technologies, customers, or application areas, I propose that **reflecting on analogous business models and prior lessons can accelerate the formulation and application of optimal strategies, preventing repeated errors and inefficiencies.**

Guided by my professional background, I formulated these business models around **Michael Porter's competitive strategy**, which has long functioned as an indispensable frame of reference in shaping industry relationships—spanning customers, internal personnel, and external partners. Yet, throughout my career, I also recognized the need to augment Porter's original model so that it more accurately reflects the complex supply chains, collaborative ecosystem, and accelerated technological innovations characteristic of the semiconductor industry. Consequently, **I introduce an Extended 5 Forces Theory, allowing a multi-layered analysis that stretches beyond primary customers to include secondary customers and the surrounding ecosystem dynamics.** By accounting for forces experienced not only by direct customers but by **"the customers of your customers,"** this expanded framework provides a more complete view of the competitive dynamics facing semiconductor firms. Put differently, the extended 5 forces framework grants companies a broader understanding of the market and bolsters their strategic agility.

Accordingly, this book applies Michael Porter's competitive strategy to semiconductors, proposing a new, integrated framework anchored in the extended 5 forces theory. Moreover, in alignment with Porter's conceptualization of competition, it proposes that the semiconductor industry is transitioning toward **'market diversification competition'—encompassing strategies for customer, application, and product diversification.**

Using these two theoretical cornerstones, I draw upon both wide-ranging industry experience and advanced insights into novel industry trends to analyze **10 revealing models**, from which I derive distinctive semiconductor business frameworks and present **10 foundational principles**.

Who is This Book for?

The central audience for this text comprises **university and graduate students** specializing in semiconductor technology and business; it is also intended for **professionals** already active in the industry. My ambition is to enable students to appreciate **how semiconductor technology connects with real-world business needs**, reflecting the interplay of semiconductor companies, their customers, industry partners, and the customers of their customers. This approach clarifies how advanced technology translates to commercial value.

Some students will commence their careers directly in a semiconductor firm's commercial division, others might start as engineers, only later assuming business or managerial roles. Regardless of their initial path, individuals who eventually ascend to broader leadership must develop strong business acumen.

Hence, **this book endeavors to sow the seeds of business consciousness for such students**, encouraging them to contemplate not only technical R&D but also the ways in which products generate revenue and profit—and ultimately serve the broader goals of humanity. Put succinctly, the fundamental mission here is to help students cultivate deeper, more holistic thinking about the semiconductor domain.

Moreover, for **professionals** currently employed in the semiconductor sector, the models and strategies outlined herein represent a toolkit that can be applied to ongoing projects or new technological developments. By systematically integrating or adapting the ideas in this text, semiconductor specialists at all levels can refine critical decisions and better manage the inherent complexities of a multifaceted business environment.

What Makes This Book Different?

This work expands on Michael Porter's 5 forces theory and adapts it expressly to the semiconductor realm, introducing the extended 5 forces theory. According to this approach, grasping semiconductor competition requires analyzing not only primary customers but also secondary customers, accompanied by the broader semiconductor ecosystem dynamics. Furthermore, drawing on Porter's concept of competition, I point out that competition in this industry is progressing toward market diversification. Expressed more explicitly, it is through **the extended 5 forces theory** and **market diversification competition** that this text examines **10 semiconductor business models**, ultimately extracting **10 critical principles**.

Chapter 1

What Does Business Mean in the Semiconductor Industry?

Chapter 1

What Does Business Mean in the Semiconductor Industry?

We begin with a simple yet fundamental query. **"What is the nature and role of business in the semiconductor industry?"** Conventionally, business is defined as the pursuit of profit through the provision of products or services to customers, and semiconductors largely conform to this paradigm. Yet here, I adopt a broader, more holistic perspective by stressing how science, engineering, and management intersect to shape the world we inhabit—a perspective that proves especially salient for semiconductors.

As shown in **Table 1**, science clarifies the principles governing our universe, engineering transforms these principles into tangible devices or offerings, and management then drives commercialization to yield profit and spark further expansion.

Table 1 : Business as Convergence of Science, Engineering, and Management

	Science	→	Engineering	→	Management
Interest (Goal)	The principle of the world	Development	Developing products	Commercialization	Profit and Growth
	Business				
Albert Einstein	◎	Refrigerator patent	○	-	X
Steve Jobs	△	-	◎	Apple Pixar	◎
Elon Musk	◎	Multiplanetary life Sustainable energy	◎	SpaceX Tesla	◎

* Relevance : ◎ Strong, ○ Medium, △ Weak, X None

Of course, real-world progress does not always unfold along neat, sequential lines: managerial imperatives can prompt new synergies between science and engineering, while breakthroughs in engineering can spur fresh avenues of scientific investigation. Overall, the synergy among these three spheres fosters the evolution of humankind's capabilities—**a synergy that underpins the semiconductor industry's structure**. Whether engaged in pure research, product design, or strategic leadership, participants in the semiconductor field invariably address business concerns that revolve around generating economic and societal value. **These business challenges** are addressed through **insights into business**

models rooted in patterns and principles, ultimately leading to **the creation of stories that shape the future**.

Examining **Table 1** more explicitly.

✓ **Science** focuses on unraveling universal laws—whether subatomic phenomena or macro scale behaviors. In semiconductors, this might entail exploring quantum mechanics or advanced materials.

✓ **Engineering** applies this scientific insight to build real products, such as new semiconductor devices or advanced integrated circuits, aiming to address practical demands. This leads to the **Development** stage, where theoretical concepts transform into prototypes or refined solutions that can then be adopted by end users.

✓ **Management** then commercializes these products and propels them into mass production, generating profit and driving further enterprise growth.

Additionally, **Table 1** includes a subjective illustration of how three individuals who substantially impacted humanity—**Albert Einstein, Steve Jobs**, and **Elon Musk**—connect to science, engineering, and management, respectively. Although somewhat symbolic, this alignment underscores how specialized contributions in one domain may guide or catalyze further innovation in the others.

While adopting this broad perspective on business is valuable, profit generation remains indispensable. Hence, I present **three fundamental formulas** relevant to profit in the semiconductor industry

1. **Profit = Price - Cost**

→ Maximizing profit depends on boosting perceived value so as to elevate price or optimizing production to reduce costs. In semiconductors, it might mean adopting advanced packaging that justifies premium pricing or streamlining processes in the FAB to keep manufacturing costs in check.

2. Definition of Value to Customers

Value = Benefit - Needs

→ The difference between a solution's benefits and the customer's underlying needs forms the basis of perceived value. Surpassing a customer's minimal functional requirements can significantly raise willingness to pay.

3. Contribution Through Value

Contribution* = ∫ *(E*, V*)

* *Experience = Knowledge x Action,* * *V = Value*

→ Customers build their "experience" of a product by combining knowledge (understanding the product's capabilities) with action (actually purchasing or employing it). As these experiences accumulate, perceived value crystallizes, leading to genuine societal or commercial contribution.

Chapter 2

Extending Porter's Strategy: New Framework Proposals

Chapter 2

Extending Porter's Strategy: New Framework Proposals

In the previous chapter, we surveyed the essence of business within the semiconductor world. Now, in Chapter 2, we reinterpret Michael Porter's competitive strategy with an eye toward semiconductor-specific traits, thus constructing the theoretical scaffolding for analyzing and deriving business models.

We begin by summarizing Porter's fundamental ideas, then extend them to the semiconductor field through the extended 5 forces theory and the evolved competition theory, both of which will inform our subsequent discussion of semiconductor business models in Chapter 3.

2.1 Proposal of the Extended 5 Forces Theory

Porter's model maintains that companies compete not merely with direct rivals but rather within a framework shaped by five structural forces. These 5 forces delineate the core economic architecture of an industry and outline how competition transpires (**Figure 1**).

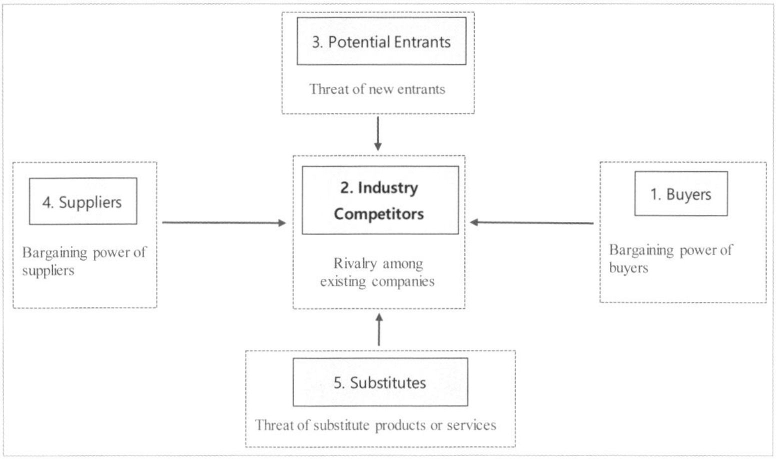

Figure 1 : Michael Porter's 5 forces theory

Summary of Michael Porter's 5 Forces Theory

1. Five Factors of the Underlying Economic Structure

- ✓ **Buyers**: The firms or individuals purchasing goods or services.
- ✓ **Industry Competitors**: The established companies vying for market share **within a particular sector** (e.g., memory or logic device producers).
- ✓ **Potential Entrants**: Newcomers who may disrupt established markets if they surmount industry barriers.
- ✓ **Suppliers**: Providers of critical resources—such as raw

materials, equipment, or contract manufacturing services—necessary for production.

- ✓ **Substitutes**: Alternative products or technologies that can fulfill the same end need.

2. 5 Competitive Forces = 5 Forces

Each factor exerts a distinct competitive pressure—shaping profitability, rivalry intensity, and strategic positioning across the industry.

- ✓ **Bargaining Power of Buyers:** To what extent can customers push down pricing or demand advantageous terms?
- ✓ **Rivalry Among Existing Competitors:** How fiercely do incumbents battle over market share and profit?
- ✓ **Threat of New Entrants:** What is the likelihood that fresh players will join the industry, fueling new rounds of competition?
- ✓ **Bargaining Power of Suppliers:** How strongly can suppliers impact price, quality, or supply conditions?
- ✓ **Threat of Substitutes:** Could alternative solutions bypass or replace the product's function, diminishing existing demand?

3. Strength of the Competitive Forces

The intensity of competition faced by a firm stems from how these five forces coalesce.

4. Summary

By applying the 5 forces, companies can map opportunities, threats, and potential profit zones, shaping strategies to respond accordingly.

Porter's 5 Forces Theory in the Semiconductor Industry

While Porter's model can be applied to nearly any industry, the semiconductor sector exhibits particular features that clarify how these five forces manifest, including intricate global supply chains, rapid cyclical demand swings, and extraordinary technology intensiveness.

Table 2 : Porter's 5 forces theory in the semiconductor industry

No.	5 Forces			Semiconductor industry
1	Buyers	Customers	**Bargaining power of buyers**	Set and Module makers, Fabless
2	Industry Competitors	+ **Your Company**	**Rivalry among existing companies**	IDM, Fabless, Foundry, OSAT, IP
3	Potential Entrants	New Competitors	**Threat of new entrants**	
4	Suppliers	Vendors	**Bargaining power of suppliers**	Materials, Equipment, Parts, IP

No.	5 Forces			Semiconductor industry
5	Substitutes	Alternatives	**Threat of substitute products or services**	New Materials, Quantum Computing

1. Buyers – Bargaining Power of Buyers

In semiconductors, key buyers include major set manufacturers (producing devices like smartphones or PCs), module assemblers, and fabless companies such as NVIDIA or AMD that outsource actual wafer fabrication. Their influence depends heavily on volume, technical expertise, and brand strength. A large buyer can demand more stringent pricing or special product features, while a smaller, technically less sophisticated customer might lack negotiating muscle.

Alternatively, A Co. (famous for a marquee smartphone line) initially sourced APs from Samsung but soon recognized how integral AP performance and branding were—thus deciding to develop its own AP design in-house, outsourcing the foundry work to T Co. This underscores how customers capable of vertical integration or advanced design expertise can pivot away from suppliers, exhibiting stronger bargaining power.

2. Industry Competitors – Rivalry Among Existing Companies

The semiconductor arena includes IDMs (like Intel, Samsung, or Micron), specialized fabless houses (e.g., Qualcomm), OSAT providers (e.g., ASE or Amkor), and IP vendors (e.g., ARM). Each type competes for market share, revenue and profitability, scale, and cost/performance advantages. Technological innovation often intensifies this rivalry, as companies race to pioneer next-generation nodes or advanced packaging.

3. Potential Entrants – Threat of New Entrants

Although entry barriers—notably massive capital expenses and proprietary knowledge—are high, new players might still emerge through alliances or niche breakthroughs. For example, certain tech giants might allocate billions to in-house silicon design, or a new foundry might rise in a region backed by heavy government incentives. Existing firms must thus maintain ongoing R&D, process efficiency, and vigilant monitoring of emergent rival strategies.

4. Suppliers – Bargaining Power of Suppliers

Critical resources—like leading-edge lithography equipment, specialized wafers, advanced EDA tools, IP cores and foundry service—are supplied by a select set of top-tier vendors. High reliance on these specialized resources can make or break a semiconductor firm's cost structure or technology road map. Effective supply chain management thus becomes a make-or-break necessity, particularly for smaller players, to mitigate the risk that a single supplier might impose unfavorable terms or raise costs unexpectedly.

5. Substitutes – Threat of Substitute Products or Services

Rapid leaps in IT and materials science can yield new solutions that challenge established semiconductor devices. For instance, a transition from conventional planar NAND to 3D NAND disrupted earlier product lines, while emerging memory (e.g., MRAM or RRAM) might eventually displace standard DRAM solutions. Meanwhile, advanced integrated SoCs can diminish demand for discrete chip solutions. Continued R&D investment helps stymie these substitute threats, sustaining a firm's competitiveness.

In sum, the 5 forces approach provides a lucid organizational structure for interpreting semiconductor industry competition. **Yet the sector's growing complexity—including advanced process nodes and lengthening supply chains—calls for a sharper lens. Thus, I propose the Extended 5 Forces Theory** that accommodates the multiplicity of specialized roles, complex ecosystem interactions, and rapid technology cycles distinguishing semiconductors.

Proposal of the Extended 5 Forces Theory for Semiconductors

Because the traditional Five Forces approach consistently guided many strategic decisions in my professional work—especially around customers, internal relationships, and partnerships—I observed first-hand the necessity to refine the model in response to the sector's rapidly maturing supply chains and accelerating technological thresholds. Thus, I propose the extended 5 forces theory to address these distinctive complexities.

In essence, **the Extended 5 Forces Theory affords a more layered vantage point, acknowledging not merely primary customers but also secondary customers, as well as the comprehensive semiconductor ecosystem dynamics. By factoring in the demands faced by the "customers' customers," we gain insight into how competition evolves across multiple tiers.** In an industry grounded in intricate technology specialization and dynamic B2B structures, failing to include the second or even third layer of the chain can obscure crucial competitive indicators, resulting in suboptimal or incomplete strategies.

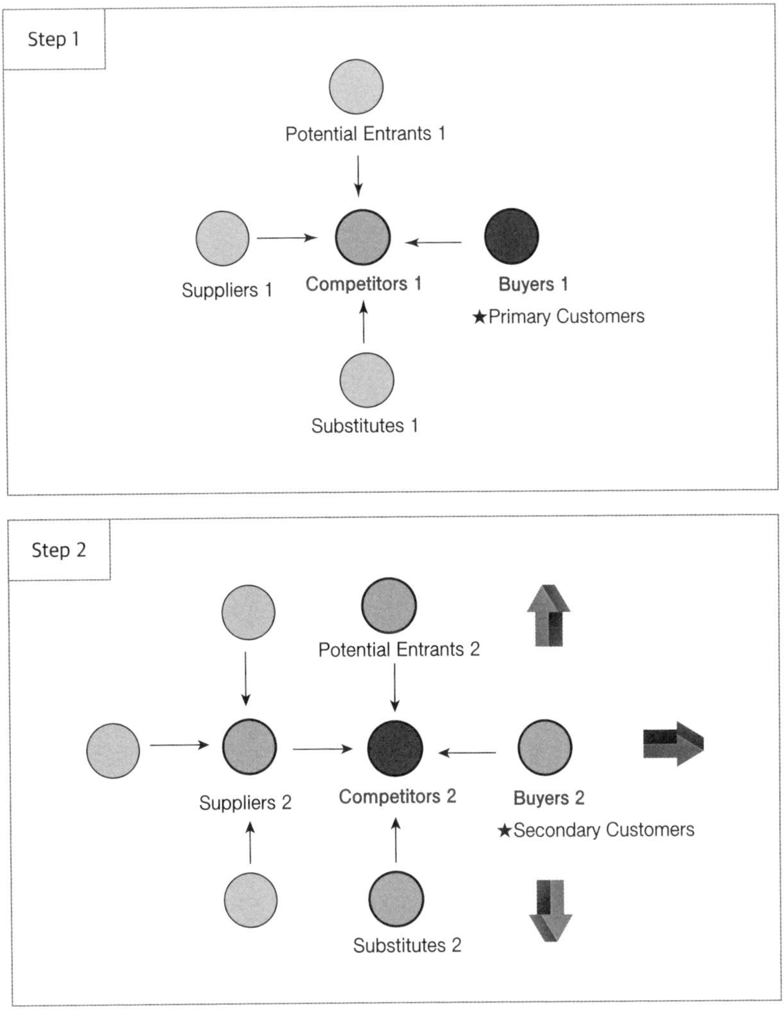

Figure 2 : The extended 5 forces theory framework (Step 2)

Step 1: Porter's 5 Forces focus on Buyers 1 (primary customers)

Step 2: The extended 5 forces theory broadens to include **Buyers 2** (secondary customers), capturing the pressure and interactions from a deeper slice of the supply chain.

When adopting **the extended 5 forces (Step2)**, the firm can evaluate,

- ✓ **Competitors 2**: Formerly Buyers 1, but now recast as direct product/service suppliers to Buyers 2.
- ✓ **Buyers 2 (the secondary customers)**: The "customers of our customers." Even if they do not directly transact with semiconductor suppliers, they can substantially influence business viability.
- ✓ **Suppliers 2:** Entities that appear as Competitors 1 in Step 1, but from the vantage of Buyers 2, they become "suppliers." This reclassification underscores the complexity of roles in B2B contexts.
- ✓ **Substitutes 2:** Alternative solutions or products that Buyers 2 might consider, replacing existing purchasing decisions.
- ✓ **Potential Entrants 2:** Prospective entrants relevant to the secondary customer's market, lowering entry barriers and stoking fresh rivalry.

The extended 5 forces structure, especially critical in B2B-based semiconductors, enables a robust recognition of how decisions by the ultimate end market or consumer can ripple back through multiple supplier layers. This multi-faceted angle consistently proves valuable when forecasting demand for newly introduced functionalities, comprehending the real impetus behind device commoditization, or anticipating next-generation design mandates.

Hence, "Who do semiconductor companies compete with?" The answer covers not only direct rivals in "their" industry but also indirect forces and shifting preferences further down the chain. In the subsequent section, we elucidate how this broader perspective fosters a deeper understanding of competition's ongoing evolution in semiconductors.

2.2 Market Diversification Competition

As previously noted, a firm's competitive landscape draws from more than just direct competitor clashes; it also arises from shifting balances of power with customers, suppliers, new entrants, and substitutes. Accordingly, **in semiconductors, I posit that competition unfolds through three stages (Table 3)**,

- ✓ **Traditional Competition:** Classic rivalries centered on "existing players," measured by metrics like share, revenue, or profitability within a stable customer base.
- ✓ **Extended Competition:** As outlined by Porter, which integrates the entire 5 forces scope.
- ✓ **Market Diversification Competition:** A distinctive phenomenon for semiconductors, in which the industry's cyclical nature and fresh technological frontiers (e.g., AI,

autonomous vehicles) demand continuous diversification—in customers, applications, and product lines.

Table 3 : The Evolution of Competition in the semiconductor industry

	Competition →	Extended Competition →	Market Diversification Competition
Competitive Players	The established players	5 Forces	Customer, Application, Product Diversification
Competitive Dynamics	Market share, Revenue, Profit	Barriers to Entry, Bargaining Power, Threat of Substitutes	Diversification effect - Enhance Competitiveness - Risk Mitigation
Strategic Framework	Competitive strategy by Porter		New proposal for Semiconductors

Amid the maturity of core sectors such as smartphones or PCs, companies scramble to diversify their activities. New application fields, especially AI or IoT, can expand or contract unpredictably, intensifying competition. Indeed, semiconductor firms must manage a delicate balance, preserving their core revenues while exploring supplementary product lines or forging alliances in uncharted application domains. If they fail to evolve, they risk stagnation or losing ground to more adaptive rivals. In response, we now explore **the proposed strategies** required for each element of this evolving **market diversification competition** in semiconductor ecosystem dynamics. Accordingly, the next section introduces proposed strategies—beginning with customer diversification.

Market Diversification Strategy

1. Customer Diversification

In the contemporary semiconductor ecosystem, customer diversification serves as a critical hedge against cyclical downturns or unforeseen disruptions. Major set manufacturers, for instance, may drastically reduce orders if consumer demand for a flagship device falters. Without a diversified portfolio of customers, a supplier could suffer catastrophic revenue swings.

Additionally, the strategic impetus toward diversification often involves establishing robust commercial relationships with **new players** in adjacent markets, such as emerging fabless startups focusing on AI accelerators. While large-scale customers typically deliver stable volumes, smaller or niche customers can offer higher margins or faster design cycles. By refining existing, mass-produced semiconductor designs—adjusting pin configurations, firmware, or packaging—companies can adapt quickly without incurring the overhead of full-scale design overhauls.

Furthermore, a well-executed customer diversification scheme typically encompasses two complementary goals:

- ✓ **Preserving** or increasing share with one's mainstay accounts, often achieved through competitive pricing, close engineering support, or advanced product road maps—serves as the foundation for successful customer diversification.

✓ **Extending** into one or more new or second-tier clients that offer incremental growth and valuable co-development experience, despite having initially limited product lines.

2. Product Diversification

In practice, **product diversification** frequently unfolds when a semiconductor supplier proves its competence by delivering a **core chip**—for example, a main SoC or memory solution—then proposes related, **secondary chips** such as power management ICs (PMIC), high-speed interface modules, or custom sensors. This layered approach fosters deeper integration into the customer's BOM (Bill of Materials) and design cycles. As the supplier's devices become integral to multiple subsystems, the supplier gains leverage to cross-sell or secure stable multi-year contracts.

Crucially, product diversification also helps address changing demands. For instance, if demand for the core chip dips due to external market conditions, the supplier can still sustain a strong relationship by selling expansion products. Over time, a well-diversified product portfolio can create "**stickiness**," making it harder for competitors to displace the incumbent once multiple chips or modules are integrated into the customer's product design and supply chain.

However, product diversification imposes new challenges, including ensuring each additional solution meets the same or higher reliability standards. The cost of R&D for expanded lines, especially if they deviate from the supplier's primary process or node, must be carefully balanced against prospective business returns.

3. Application Diversification

Application diversification stands at the forefront of long-term semiconductor strategies. While companies may excel in consumer electronics—like smartphone DRAM or SoCs—they often seek entry into automotive, industrial, or data-center markets. Yet each domain demands a distinct set of standards: automotive semiconductors require extremely robust **AEC-Q100** reliability standards, extended temperature ranges, and multi-year support cycles; industrial IoT chips might need specialized security features or real-time performance.

The central barrier is usually R&D cost, migrating from a smartphone-optimized design to an automotive or industrial grade device requires significant modifications in packaging, fault tolerance, and test procedures. Time to market can stretch out, and automotive qualification, for instance, might take an extra 12 to 24 months. Meanwhile, well-entrenched competitors already serving these markets enjoy brand trust and validated track records.

Hence, **a measured approach—identifying synergy between**

existing design IP or manufacturing processes and new industry requirements—typically offers the best route. Even so, achieving breakthroughs in **EV**(electric vehicle) power electronics or **edge AI** sensors can deliver substantial growth, offsetting the risk if a smartphone or consumer segment dips.

Moreover, application diversification mitigates large-scale cyclical volatility; if one sector slows (e.g., consumer electronics), other emerging markets (e.g., automotive sensor arrays or industrial automation) might sustain or even accelerate demand. **This resiliency is arguably the biggest strategic gain from application diversification**.

Strategic Implications of Market Diversification

From an **academic** standpoint, market diversification can be interpreted in light of **Ansoff's Matrix**, which categorizes growth strategies according to whether a firm pursues new/existing products and new/existing markets. In semiconductors, each move—be it from consumer SoC to automotive-grade SoC or from domestic handset OEMs to foreign industrial clientele—can be framed as an extension of that matrix. The essential premise is that **diversification fosters resilience and optionality**, yet it also demands **advanced resource planning** (manufacturing capacity, R&D scheduling) to avoid spreading engineering teams too thin.

In an industry marked by intense process-node competition (e.g., TSMC, Samsung Foundry, Intel Foundry Services) and swift product life cycles, the capacity to pivot from one sector's slowdown to another's growth is valuable. Failing to do so can see a once-prosperous memory vendor or consumer SoC supplier overshadowed by a competitor that invests in emergent AI modules, high-performance computing, or ADAS chips. Hence, thorough **internal audits** of IP reuse, packaging design compatibility, or reliability can direct which diversification paths are not only feasible but also yield synergy with existing capabilities, thereby maximizing the likelihood of a profitable outcome.

2.3 Framework for Competitor and Company Analysis

In essence, Porter's framework (Figure 3) posits that understanding how a **competitor (or one's own firm)** envisions its future goals, current strategies, critical assumptions, and overall capabilities can shed crucial light on likely decisions and potential shifts. This analysis can be reversed for **self-reflection**: **a semiconductor company** can systematically evaluate its own readiness in each domain, clarifying if it lags behind in core areas like technological innovation, advanced node yield, or market diversification.

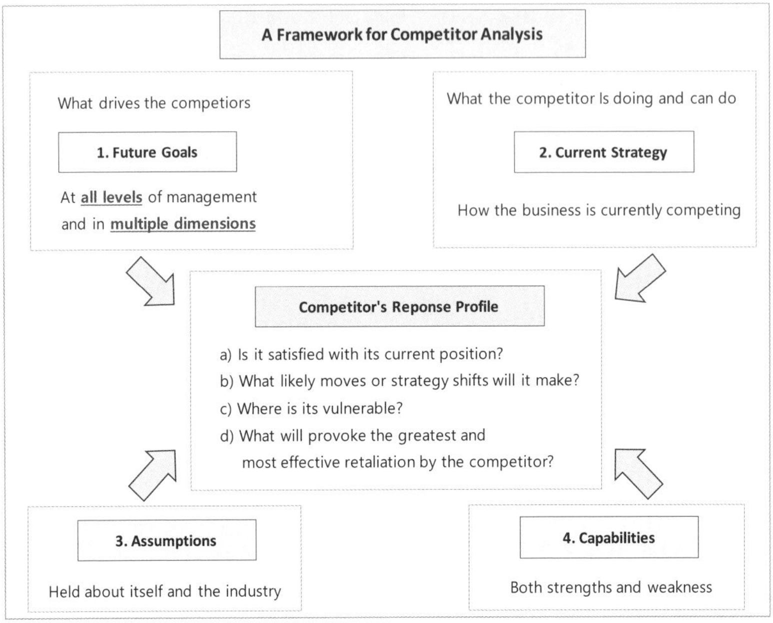

Figure 3 : Michael Porter's framework for competitor analysis

1. Future Goals

Under "Future Goals," we might see a competitor planning to adopt advanced sub-3 nm nodes, or aiming to expand from standard DRAM into specialized HBM for AI. Alternatively, a design house might schedule a road map for next-generation GPU designs that incorporate novel interconnect architectures—all as part of their **broader pursuit of technological innovation**. Any such forward-looking plan dictates resource allocations and partnerships, forming the crux of a competitor's forward path.

Meanwhile, from an external vantage, the semiconductor

company can glean how soon the competitor expects to achieve certain yield thresholds or cost reductions. This can shape one's own planning—for instance, if a competitor's future goals revolve around capturing automotive markets, a well-timed pivot to automotive-rated chips could help preempt them.

2. Current Strategy

The "current strategy" often emerges in **present-day** areas like a competitor's marketing angle (cost vs. performance leadership), its choice of foundry partners, or alliances with major IP houses. For instance, a competitor dedicated to advanced AI SoCs might partner with TSMC for leading-edge processes at 3 nm or below. Alternatively, an analog/mixed-signal chip maker might prioritize specialty processes in automotive or industrial realms. Observing these actions clarifies the competitor's immediate goals, guiding potential countermoves, collaborative or otherwise.

Fundamentally, **Yield** optimization remains the bedrock: successful semiconductors must achieve stable, acceptable yields quickly enough to command a viable margin. A competitor pushing advanced nodes prematurely might endure poor yields, eroding its cost advantage or hamper early product availability. By systematically tracking yield ramp rates, a company can anticipate if a competitor's "current strategy" is threatened by insufficient process maturity or vice versa.

3. Assumptions

In semiconductors, these "assumptions" can encompass beliefs about consumer electronics growth continuing unabated, or that HPC/AI demand for specialized HPC DRAM will remain robust for another decade. They might also reflect broader macro assumptions (e.g., about foundry capacity expansions in certain regions or about the global memory cycle soon reaching a supply shortage). If a competitor assumes "Moore's Law is plateauing," it might shift R&D to advanced packaging like 2.5D/3D integration. Contrastingly, a competitor confident in the continued viability of scaling might double-down on advanced photo lithography like high-NA EUV.

Discerning these assumptions helps a company foresee if the rival might,

- ✓ Launch new product lines earlier or later than expected,
- ✓ Re-price certain offerings in anticipation of supply constraints,
- ✓ Align closely with certain node transitions or EDA tools,
- ✓ Seek acquisitions or mergers to strengthen IP capacity in a domain the rival considers vital.

4. Capabilities

Capabilities might be dissected into,

- ✓ **Design competence**: advanced SoC design flows, specialized analog/mixed-signal expertise, proprietary AI architectures, or HPC memory design knowledge—often underpinned by effective IP or EDA tool usage.

- ✓ **Manufacturing or packaging** skill: e.g., proven mastery of advanced nodes or advanced interposer-based packaging that can significantly differentiate performance.
- ✓ **Operational proficiency**: yield learning curves, supply chain management, risk mitigation, synergy with OSATs (Outsourced Semiconductor Assembly and Test) or second-sourcing strategies.
- ✓ **Marketing, business, and brand strength:** intangible but still fundamental, especially if the competitor enjoys brand recognition in certain verticals like automotive or HPC.

In short, Porter's competitor analysis provides a guiding method for a semiconductor company to systematically evaluate **both the external threats posed by competitors and its own internal readiness**. By merging it with the concept of market diversification, a firm obtains a thorough vantage on how to prioritize R&D, approach new customer segments, or protect existing ones. Because semiconductors revolve around advanced, capital-intensive technology, neglecting either dimension—technological innovation and market diversification—can hamper long-term survivability. Thus the key takeaway from **Table 4**, summarizing the above, is that **technological innovation** and **market diversification** are both essential.

Table 4 : Framework for Competitor and Company Analysis to the Semiconductor Industry

No.	Component	Semiconductor Industry	
		Objective	Definition
1	**Future Goals**	Technological innovation	Advancing processes, architecture, new chip technologies
		Market share expansion	Increasing sales and dominance in the markets
		Customer, Application diversification	Securing new clients across various industries
		Capacity expansion	Expanding production leading the economies of scale
2	**Current Strategy**	Successful transition from research to development	Technical feasibility, Economic viability, Supply chain investment
		Enhancing yield	New product quality, Productivity and Supply Stability, Profitability
		Strengthening partnerships with key customers	Business continuity
		Product diversification	Risk mitigation, Leveraging existing resources
3	**Assumptions**	Beliefs and assumptions	For itself, Competitors, Circumstances (3C)
4	**Capabilities**	Both strengths & weakness	Resources, skills, and technical expertise

Chapter 3

Semiconductor Business Models and Strategies

Chapter 3

Semiconductor Business Models and Strategies

Having established our theoretical base—Porter's approach, refined via **the extended 5 forces** and an evolving sense of **market diversification**—we now shift focus to an in-depth examination of **10 distinct semiconductor business models**. These models, rooted in direct or indirect experiences within the industry, provide a platform for extracting **10 key principles** relevant to understanding tomorrow's semiconductor products, markets, and relationships.

The primary objective is to equip semiconductor professionals, students, and leaders with conceptual tools to interpret existing patterns and apply them to next-generation solutions. Each model is interpreted through the lens of both extended 5 forces and market diversification, thereby exposing intricate relationships that might otherwise remain overlooked.

Model 1.
Secondary Customer Market-Driven Diversification

1) Business Model Analysis Through the Extended 5 Forces

This first model deals with the supply of display modules and related semiconductors in consumer-friendly tablet devices

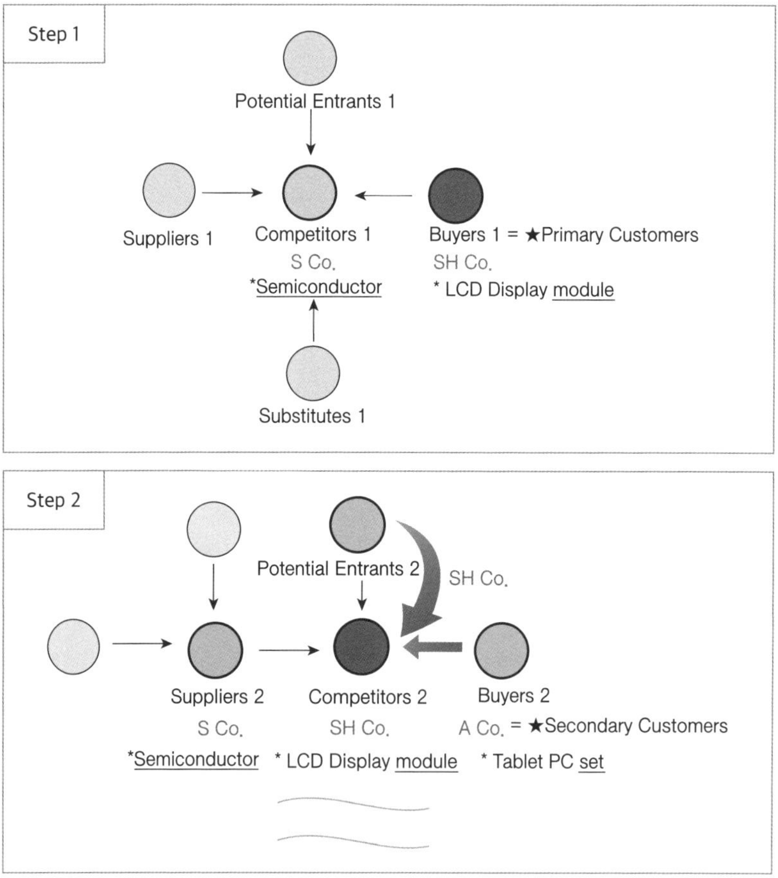

Figure 4 : Extended 5 forces framework for model 1

(like iPads). Although the final products occupy mainstream consumer markets, the business behind them involves sophisticated supply chain interplay, particularly among module houses and semiconductor suppliers.

Figure 4 illustrates **the extended 5 forces framework** for Model 1, underlining how the business environment can change substantially when we consider not only direct (primary) customers but also **secondary** ones. In traditional analyses (**Step 1**, Porter's 5 forces), the semiconductor company S Co. is shown to supply a device (e.g., a display driver) to SH Co., a display module manufacturer. While the usual 5 forces approach might suffice to outline competition and bargaining power at that first level, it offers only a partial view of what eventually propelled S Co. to secure the order.

By adopting **Step 2** (the extended 5 forces theory), we shift the lens to **incorporate the "secondary" or "downstream" customer,** referred to here as **A Co**. (Buyers 2). In this scenario, A Co.—the set manufacturer of a highly popular tablet—opted to diversify its module suppliers (Competitors 2) for cost optimization and improved supply stability. Hence, A Co. actively added **SH Co**. (formerly just a potential entrant for A Co.) to its official list of approved module vendors.

S Co. seized the opportunity to promote its existing, mass-produced semiconductor device—an already field-proven solution—to **SH Co.** This device had been successfully supplied to another module maker, thus conferring it a reliable **mass-production record** and proven functionality. By highlighting those attributes and making minimal engineering modifications, S Co. demonstrated that it could fulfill SH Co.'s needs with minimal time-to-market. This timely approach proved pivotal in winning the order.

In essence, the "secondary" customer, A Co., possessed strong bargaining power—enough to dictate that multiple module vendors compete for its tablet display orders. That impetus meant the new or "secondary" module company (SH Co.) needed to secure a semiconductor supplier quickly, thereby creating the ideal opening for S Co. In effect, the extended 5 forces framework shows how the secondary customer's preferences can reshape a market structure **two steps upstream,** enabling new alliances and challenging incumbents who do not respond swiftly.

Strategic Significance of Secondary Customer Power

This example underscores the case where a secondary customer (A Co.) wields substantial market leverage, mandating diversification among its first-tier suppliers (the display module makers). **By carefully analyzing not just direct customers, but also the strategic moves of the customers' customers, a semiconductor firm can position itself to exploit new openings**. More specifically, S Co.'s success hinged on:

- ✓ **Monitoring** any shift in the secondary customer's supply strategy.
- ✓ **Promoting** the product's proven track record, reassuring SH Co. of fast qualification and stable yields
- ✓ **Adapting** an existing, mass-produced device with only small modifications (thus minimizing cost, schedule, and risk).

2) Key Force and Market Diversification Strategies

When we examine the extended 5 forces, it becomes apparent that **the bargaining power of the secondary customer** is the key competitive force driving this model. We label this structure **"Secondary Customer Market-Driven Diversification."** The secondary customers, in search of diversified supply lines, effectively forced the primary customers (module) to expand their vendor lists—and this chain reaction opened the door for S Co. to become an approved semiconductor supplier to SH Co.

Hence, semiconductor providers can improve their chances of obtaining new business by studying the secondary (and even tertiary) layers of their supply chain. Once the secondary customer decides to add or swap a major supplier, it can translate directly into opportunities at the semiconductor level. By understanding the extended 5 forces, S Co. aligned itself with these shifts, thereby cultivating a new relationship that ultimately expanded its direct customer portfolio.

Table 5 : Extended 5 forces and market diversification factors for model 1

Component	Factor		Business Model
Extended 5 Forces Theory	Key Force	Bargaining power of secondary customers (Buyers 2)	◎
	Strength of the Competitive Forces for Semiconductor		○
Suppliers 2 analysis (Semiconductors, S Co.)	**Technological innovation**	Technological innovation	△
		Successful transition from research to development	△
		Enhancing yield	△
	Diversification	Customer	◎
		Application	△
		Product	○

* Relevance: ◎ Strong, ○ Medium, △ Weak

Table 5 delineates how **the extended 5 forces** and **market diversification** interact for Model 1. From S Co.'s perspective (as Suppliers 2), the core strategy was to adapt a **"mass-produced product"** for SH Co. with only modest technical innovation. This approach:

- ✓ **Preserved** S Co.'s ties to its existing main customers.
- ✓ **Gained** a new customer (SH Co.) under minimal additional R&D burden.
- ✓ **Fulfilled** A Co.'s secondary-customer-driven expansion plan.

This synergy exemplifies **customer diversification** at its best: S Co. expanded its base to include SH Co. while retaining profitable relationships with existing partners. It also reiterates a critical principle: the more a company can re-purpose or re-qualify a proven solution with only slight adjustments, the faster it can convert new leads into stable orders, thereby minimizing cost, engineering overhead, and time to revenue.

Reflective Question: If you reevaluate your own industry, who might be the secondary customers that inadvertently steer your future prospects? Are they large global OEMs mandating supply chain expansions? Or could they be government agencies requiring new standards that ripple back through the chain? By adopting the extended 5 forces viewpoint, you can identify such influences earlier and craft more potent sales strategies.

Model 2.
Supplier Strategies for Diversification by Invisible Forces

1) Business Model Analysis Through the Extended 5 Forces

When analyzing semiconductors, it is simpler to diversify customers for standard commodity-like products (e.g., certain memory chips). However, for specialized or custom semiconductors (e.g., SoC)—where each design is tailored to a unique customer's SPEC—diversification demands heavier resource commitments. This second model highlights a scenario where the semiconductor maker, S Co., strove to broaden its customer base for a resource-intensive device while ensuring stable demand.

Adapting Existing Products for New Customers

A "smart" method to diversify is to adapt a mass-produced solution already in volume production for a different or earlier customer, thus limiting rework. Provided the new customer has relatively predictable or stable demand, this is often the "ideal scenario" because it supplies incremental revenue without incurring a fresh wave of R&D. Here, S Co. considered three prospective customers to expand beyond the existing partnership:

✓ **G Co.**: A global technology firm entering the mobile phone sector for the first time—large brand power but unverified manufacturing and commercialization experience.

- ✓ **B Co.**: A domestic manufacturer with rich experience but declining volumes, raising ROI concerns.
- ✓ **Multiple Chinese phone OEMs**: Potential for significant volume but questionable competition environment and uncertain ROI, given intensifying price wars and saturating phone markets.

Invisible Forces: AI and Beyond

As a result, S Co. faced a strategic decision, with the need to choose one among three potential customers. To navigate this decision, it becomes necessary to explore both conventional and emerging strategic frameworks.

Figure 5 illustrates the extended 5 forces framework for Model 2, serving as a conceptual lens through which we can analyze how both the extended 5 forces and the emerging concept of "**invisible forces**" are applied in this model. In the initial analytical stage—referred to as Step 1, which corresponds to Porter's conventional 5 forces—the focus is solely on S Co.'s supply relationship with its established mobile phone set customer, SM Co. At this level, the analysis is confined to direct interactions between supplier and primary customer, highlighting factors such as bargaining power and immediate market pressures. However, this initial level of analysis is inherently limited because it does not capture the broader strategic environment that influences future demand or potential shifts in customer behavior.

Recognizing this limitation, the framework advances to **Step 2—**

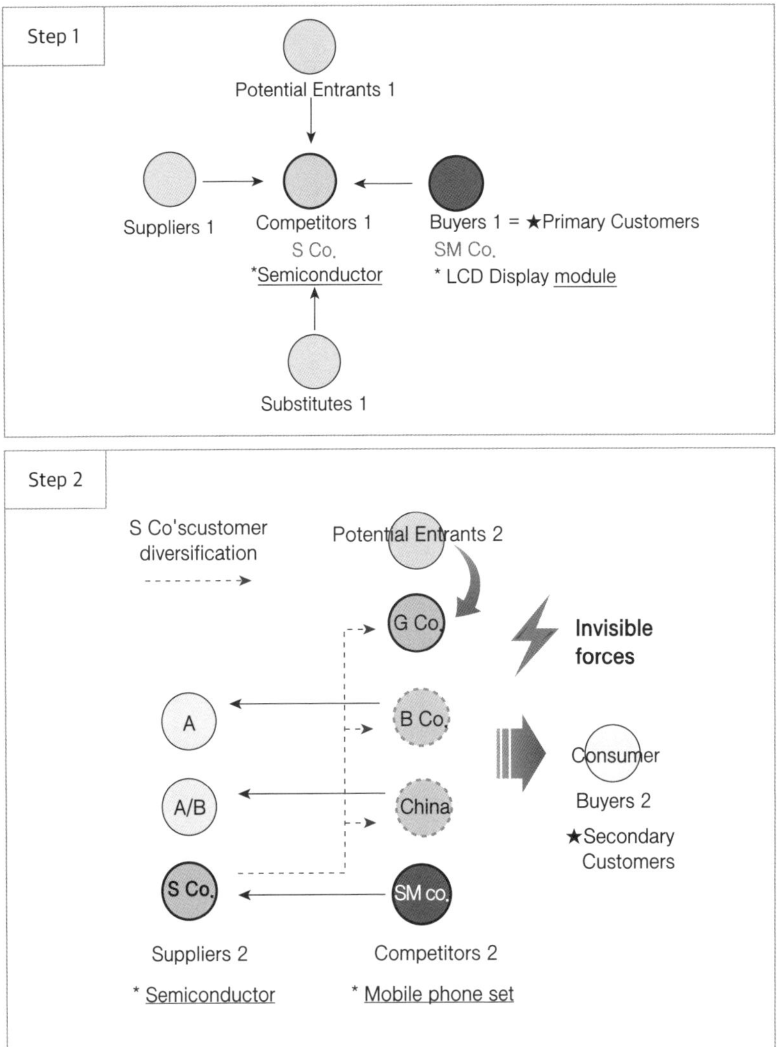

Figure 5 : Extended 5 forces framework for model 2

the extended 5 forces theory—which incorporates an additional layer by considering secondary customers (Buyers 2) and their unique influences. In this extended analysis, S Co. is no longer viewed only in terms of its relationship with SM Co., but rather, it is seen as actively evaluating prospective new customers, such as G Co., B Co., and various Chinese mobile phone companies. These prospective customers, collectively grouped as Competitors 2 in the context of the secondary market, become a critical dimension when S Co. seeks to diversify its customer base beyond its longstanding relationship with SM Co. This added layer of analysis provides a more holistic view of the market dynamics and better illuminates the latent opportunities for strategic expansion.

While this extended analysis broadens the strategic lens, it still may not fully capture all underlying drivers of change. At this juncture, the concept of **invisible forces** is introduced. These invisible forces refer to **potential factors** that are not immediately evident or directly captured by the traditional or even the extended 5 forces framework, yet **have the capacity to dramatically influence future market and competitive dynamics**.

By framing these **invisible forces as elements that orbit just beyond the extended 5 forces,** semiconductor companies are encouraged to remain vigilant and adaptive, scanning not only for immediate competitor moves but also for **emergent catalysts capable of reshaping markets.** Keeping pace with such uncertain influences demands **an openness to constant learning and iterative strategy refinement.** Hence, **invisible forces do more than occasionally disrupt the status quo—they often define the next competitive frontier.**

One notable example of an invisible force that S Co. considered was the emerging impact of AI. Four to five years ago, AI's transformative potential had not yet fully manifested in the consumer market. Nevertheless, S Co. recognized AI as a potential game-changer that could eventually redefine product requirements and market dynamics from the perspective of primary customers (i.e., set companies). Although predicting the specific influence of AI on the mobile phone market was fraught with uncertainty at that time, S Co. prudently integrated this possibility into its strategic evaluation.

By considering AI as an invisible force, S Co. eventually decided to pursue a strategic partnership with G Co. Even though G Co. was new to the mobile phone business and its production volumes were difficult to predict, the long-term prospects associated with AI adoption offered a compelling reason to invest in this relationship. Over time, this strategic decision was vindicated: B Co. eventually closed its mobile phone business amid worsening economic conditions, and Chinese companies, confronted with intensified market competition, were unable to commit further resources to complex, resource- and cost-intensive semiconductor components. Meanwhile, G Co.—with its renewed focus on AI and the corresponding need to develop and produce AI-integrated mobile phones—persisted in its business operations despite initial profitability challenges.

This model, therefore, demonstrates that for semiconductor companies, pursuing customer diversification with limited resources can be achieved effectively by incorporating the concept of invisible forces into the extended 5 forces analysis. By doing so, firms can make well-informed strategic decisions that not only mitigate immediate risks but also position them advantageously for future market shifts.

2) Key Force and Market Diversification Strategies

When we analyze this model through the lens of the extended 5 forces, it becomes apparent that the **bargaining power of Suppliers 2** emerges as the key competitive factor. In this context, Suppliers 2—represented by S Co.—must decide which new customers to invest in based on the resources required and the potential stability of future demand. For semiconductor companies undertaking customer diversification, a meticulous and resource-sensitive approach is critical. S Co. chose to modify an already mass-produced product with only minimal engineering adjustments. This strategy was implemented in a market where the growth of smartphones was beginning to slow, thereby necessitating careful resource management while still expanding into new customer segments.

Based on this comprehensive analysis, the business model is aptly named **'"Supplier Strategies for Diversification by Invisible Forces."** This title reflects the dual emphasis on both the necessity to diversify customer relationships in a challenging market environment and the importance of recognizing and preparing for invisible forces—such as the potential disruptive impact of AI, that may alter competitive dynamics in the near future.

Table 6 : Extended 5 forces and market diversification factors for model 2

<table>
<tr><th>Component</th><th colspan="2">Factor</th><th>Business Model</th></tr>
<tr><td rowspan="2">Extended 5 Forces Theory</td><td>Key Force</td><td>Bargaining power of suppliers 2</td><td>○</td></tr>
<tr><td colspan="2">Strength of the Competitive Forces for Semiconductor</td><td>○</td></tr>
<tr><td rowspan="6">Suppliers 2 analysis (Semiconductor)</td><td rowspan="3">Technological innovation</td><td>Technological innovation</td><td>△</td></tr>
<tr><td>Successful transition from research to development</td><td>△</td></tr>
<tr><td>Enhancing yield</td><td>○</td></tr>
<tr><td rowspan="3">Diversification</td><td>Customer</td><td>◎</td></tr>
<tr><td>Application</td><td>△</td></tr>
<tr><td>Product</td><td>△</td></tr>
</table>

* Relevance: ◎ Strong, ○ Medium, △ Weak

To further elucidate this model, **Table 6** outlines the extended 5 forces and market diversification factors from the perspective of Suppliers 2 (S Co.). In this table, the following key points are made,

- ✓ S Co. successfully adapted an already mass-produced product to meet the specific requirements of new prospective customers. This adaptation required only a modest level of technological innovation, thereby minimizing R&D costs while maximizing speed to market.
- ✓ In terms of market diversification, S Co. adopted a strategy that focused on securing new customers while retaining its existing base. This balanced approach is crucial for mitigating the risks associated with market concentration.

Moreover, in today's market environment—where AI technology has risen to prominence—every semiconductor company is emphasizing AI in its product differentiation strategies. Thus, it is incumbent upon semiconductor companies to perform **the extended 5 forces analysis** that not only covers the immediate supply chain of primary customers but also consistently monitors **the secondary customers' landscape.** By doing so, companies can **identify emerging invisible forces** and adjust their strategies accordingly.

In closing this section, the text invites reflection: What factors might be considered invisible forces in your industry? Such forces might include geopolitical shifts, unexpected regulatory changes, or evolving consumer preferences that subtly reshape demand curves over time. Exploring these dimensions, and effectively integrating them into your strategic planning, can yield substantial long-term advantages.

Model 3.
AI Semiconductor Ecosystem

1) Business Model Analysis through The Extended 5 Forces

Using the extended 5 forces framework, we analyze **AI semiconductors**, which constitute one of the most transformative, high-

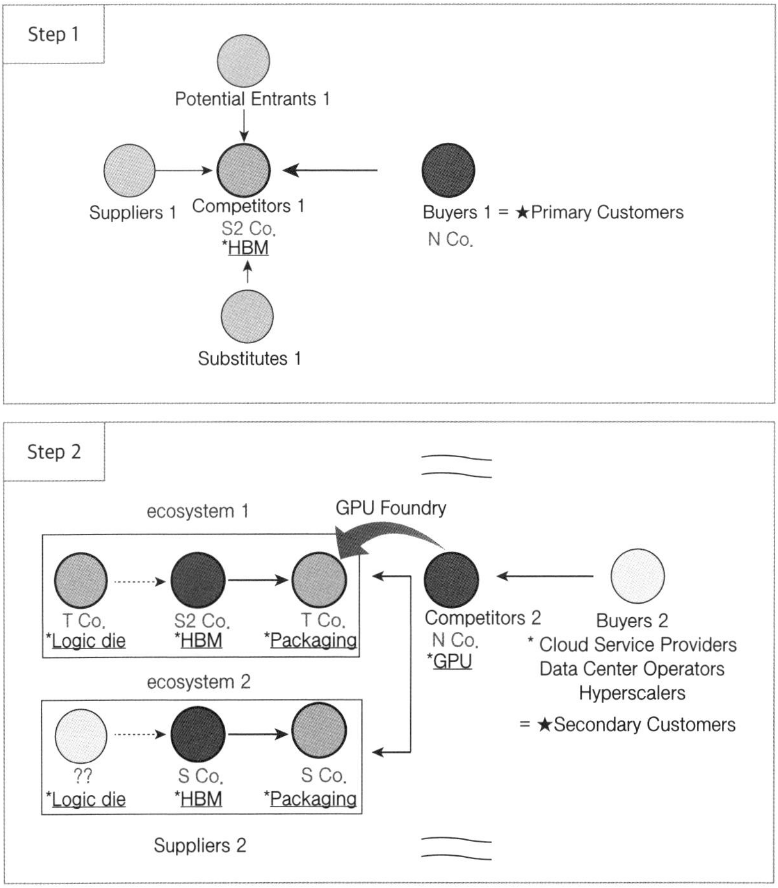

Figure 6 : Extended 5 forces framework for model 3

profile advancements in microelectronics.

Figure 6 illustrates that in **Step 1,** one can view the structure where **S2 Co.** supplies **HBM** semiconductors to **N Co.**. At this initial level, we see that the primary customers (Buyers 1)—such as N Co.—possess considerable negotiating power. Yet, restricting the analysis to Step 1 alone limits our view of how AI semiconductor systems genuinely operate. Therefore, we shift to **Step 2** of the extended 5 forces theory, which introduces **secondary customers** (Buyers 2), i.e., **cloud service providers,** data center operators, and major hyperscalers. At this downstream layer, **N Co.** repositions itself as **Competitors 2,** while Competitors 1 evolve into **Suppliers 2.**

Because AI semiconductor solutions—such as GPUs, AI accelerators, and specialized memory—demand advanced technology, manufacturing segmentation, and end-user partnerships, no single enterprise can handle every stage independently. Thus, multiple participants each develop specialized roles in the design, foundry work, packaging, or memory supply. These specialized **"Suppliers 2"** unite to form a cohesive but intricate **AI semiconductor ecosystem,** in which collaboration is essential to bring final products to market in a cost-effective, timely manner.

Origins of an Ecosystem Structure

As the text explains, the AI semiconductor sector demands a level of **complexity and performance** that strains the resources of any single enterprise. Each company chooses to refine and emphasize a core competency—for instance, advanced GPU design, state-of-the-art foundry processes, or sophisticated high-bandwidth memory (HBM). By assembling these distinctive contributions, the industry collectively produces final chips that cater to next-generation AI workloads.

More concretely,

- ✓ N Co. excels at GPU or AI logic design but relies on T Co. for advanced wafer processing at sub-5 nm nodes.
- ✓ Memory suppliers—like S2 Co. or S Co.—must deliver HBM solutions that handle massive data throughput.
- ✓ T Co. (the foundry) frequently undertakes final packaging, bonding GPUs with HBM or other specialized dies, integrating them into powerful AI modules.

Hence, **the extended 5 forces analysis** reveals that once we incorporate **Buyers 2** (cloud giants, etc.), the notion of **"Competitors vs. Suppliers vs. Buyers"** merges into an **ecosystem** where specialized roles and alliances supersede traditional boundaries. From this perspective, the AI semiconductor ecosystem reflects **two distinctive pillars—specialization and collaboration**—which are explored in the next section.

Specialization in AI

The text outlines three key pillars of specialization in AI semiconductors,

a) **Technological Expertise**

N Co. invests in GPU design or AI IP, T Co. masters advanced nodes, and memory firms perfect HBM. Each domain is so specialized that synergy is mandatory for a final product to reach top performance metrics.

b) **Manufacturing Proficiency**

AI semiconductors demand extreme yields, unique packaging (2.5D or 3D integration), and faster iteration cycles. The pace from research → development → production must be swift and efficient, with yield proving an early gate to commercial viability.

c) **Customer Responsiveness**

AI semiconductors must co-evolve with software stacks, machine learning frameworks, and HPC workloads. N Co. not only designs chips but also collaborates deeply with major cloud operators like AWS, Microsoft, or Google, ensuring that end solutions match real-world AI model training or inference requirements. By specializing in customizing HPC boards or data center platforms, N Co. reinforces its position.

Ecosystem Collaboration

Specialization permits each company to maximize core advantages, whereas collaboration ensures the entire pipeline converges into workable final products. As AI semiconductors grow more advanced, the gap between these top-tier ecosystem participants and external companies widens. Firms lacking specialized technology or alliances may struggle to break into HPC or data center markets, reinforcing the ecosystem's cohesiveness.

The AI sector also experiences **rapid obsolescence** of older nodes—when a new generation of HPC chips launches, older designs often lose competitiveness quickly. Consequently, each entity invests heavily in next-generation process nodes, more advanced memory solutions, and improved interconnect technology. This arms race further heightens the cost barrier for outsiders, crystallizing the impetus for specialized synergy.

Speed of Technological Evolution

The text underscores the reality that once new AI solutions debut, existing products rapidly become outdated, impelling every stakeholder to invest in cutting-edge processes or architectures. Given these formidable costs, we see that **economies of scale** emerge, each participant, by focusing on one slice of the pipeline, can amortize R&D across multiple HPC clients or large-run data center demands. The entire system thrives as a **communal ecosystem**, each participant fulfilling a unique but interdependent role.

2) Key Force and Market Diversification Strategies

Table 7 : Extended 5 forces and market diversification factors for model 3

<table>
<tr><th>Component</th><th colspan="2">Factor</th><th>Business Model</th></tr>
<tr><td rowspan="2">Extended 5 Forces Theory</td><td>Key Force</td><td>Rivalry among existing companies</td><td>◎</td></tr>
<tr><td colspan="2">Strength of the Competitive Forces for Semiconductor</td><td>◎</td></tr>
<tr><td rowspan="6">Suppliers 2 analysis (Semiconductors)</td><td rowspan="3">Technological innovation</td><td>Technological innovation</td><td>◎</td></tr>
<tr><td>Successful transition from research to development</td><td>◎</td></tr>
<tr><td>Enhancing yield</td><td>◎</td></tr>
<tr><td rowspan="3">Diversification</td><td>Customer</td><td>△</td></tr>
<tr><td>Application</td><td>◎</td></tr>
<tr><td>Product</td><td>◎</td></tr>
</table>

Table 7 references the extended 5 forces and market diversification factors for Model 3. From the vantage of **Suppliers 2**—in this case, HBM memory vendors—the text pinpoints **rivalry among existing companies** as the key force. Among HBM providers, advanced node transitions, yield improvements, or sub-architectural enhancements can spark fierce competition.

Further, analyzing the scene from a **market diversification** angle, we see the introduction of AI as not just a technology domain but also a strong impetus for both application and product diversification. A memory company may produce standard DRAM for PCs but then invests in specialized HBM lines specifically for

HPC or AI accelerators, broadening its product portfolio to serve a faster-growing sector.

"AI Semiconductor Ecosystem" is thus the formal name for this business model, highlighting how advanced AI chips require enormous synergy among multiple parties, each focusing on what they do best.

Case Study:
NVIDIA

NVIDIA has long recognized the transformative potential of artificial intelligence (AI) and deep learning, devoting over two decades to rigorous research and development. This steadfast commitment has positioned the company among the most respected and innovative entities in the semiconductor industry. In examining NVIDIA's journey, we derive critical insights from both a business model and strategic perspective—insights that not only illuminate the pathways of success in this highly competitive field but also serve as guiding principles for future endeavors in semiconductor innovation.

Historical Context and Technological Evolution

Since the early 2000s, NVIDIA has been at the forefront of advancing GPU technology. In 2006, the company launched CUDA (Compute Unified Device Architecture), a pioneering platform that redefined the role of GPUs. No longer confined solely to graphics processing, GPUs under CUDA quickly evolved into powerful parallel processing units capable of handling a broad spectrum of computationally intensive tasks. This breakthrough laid the groundwork for GPUs to become essential components in scientific computing, data analysis, deep learning, and ultimately, AI model training.

At this juncture, it is important to note that the development of CUDA was not an isolated technical achievement. Rather, it represented a paradigm shift in computational hardware, creating a bridge between hardware design and software innovation. This integration allowed developers to harness the massive parallelism of GPUs for general-purpose computing tasks—a concept that had far-reaching implications for industries ranging from high-performance computing (HPC) to autonomous systems. As a result, CUDA became a catalyst for the democratization of deep learning, enabling both academia and industry to experiment with complex models and large datasets.

Turning Point in AI and Deep Learning

The year 2012 marked a seminal moment for NVIDIA. During this period, the deep learning model AlexNet, which achieved a landmark victory in the ImageNet competition using NVIDIA GPUs, dramatically demonstrated the extraordinary potential of GPUs in AI research. This event not only validated the technology but also accelerated NVIDIA's strategic pivot toward deep learning and AI. Following AlexNet's success, the company intensified its investment in AI-centric R&D, which allowed it to rapidly transition from being a prominent GPU manufacturer to an essential provider of AI infrastructure.

This strategic redirection was underpinned by NVIDIA's ability to foresee the convergence of high-performance computing with artificial intelligence. The success of AlexNet functioned as a proof-of-concept that sparked widespread interest in deep learning applications. Consequently, NVIDIA leveraged this momentum to forge deep collaborations with industry titans and academic institutions alike. The company's proactive investments not only enhanced its technological portfolio but also solidified its market position in emerging sectors, such as autonomous vehicles, medical imaging, and smart robotics.

Lessons from Failure: The Tegra Experience

Despite its resounding success in the AI domain, NVIDIA's journey has not been without setbacks. A notable example is the company's Tegra series, which was an ambitious attempt to penetrate the smartphone and tablet market as part of the mobile Application Processor (AP) segment. Tegra was engineered to integrate high-performance GPUs and CPUs, and it was designed to excel in multimedia functionality. However, Tegra ultimately failed to gain traction in the mobile AP market due to a confluence of technical and market-related challenges.

Several critical issues contributed to Tegra's under performance:

- ✓ **Competitive Pressures:** NVIDIA faced fierce competition from established rivals such as Qualcomm and Samsung. Qualcomm's Snapdragon series, in particular, benefited from integrated LTE support—a feature that quickly became indispensable in modern smartphones.

✓ **Technical Limitations:** Although Tegra delivered strong graphics performance, it suffered from significant power efficiency issues. In a mobile market where battery life is a critical parameter, Tegra's high power consumption rendered it less attractive to major smartphone manufacturers.

✓ **Partnership Challenges:** NVIDIA was unable to secure partnerships with key set makers like Samsung and Apple. While Samsung favored its in-house processors and Apple was simultaneously developing its proprietary APs, Tegra was consequently sidelined in favor of solutions that aligned more closely with the strategic visions of these dominant players.

The Tegra experience, though a setback, provided **NVIDIA with invaluable strategic lessons**. The company learned that venturing into markets where it lacked core competencies—such as mobile APs—could dilute its technological focus and undermine its long-term prospects. Instead, the firm recognized the necessity of concentrating on its primary strength: GPU technology. By refocusing its efforts on developing advanced GPUs and deep learning architectures, NVIDIA was able to leverage its R&D investments and establish a robust, differentiated position in the AI space. This redirection was critical; it not only salvaged the company's competitive standing but also paved the way for long-term success by building strategic partnerships with industry leaders such as Microsoft, Google, Amazon Web Services (AWS), and Open AI.

Strategic Lessons and the Role of Time

Ultimately, NVIDIA's success story offers a compelling illustration of how a company can achieve success through a combination of **technological excellence**, **strategic focus** and **patience**. From a market diversification perspective, **the key takeaway is that concentrating on one's core strengths**—despite setbacks in ancillary markets—can lead to transformative breakthroughs. NVIDIA's eventual success in the AI market, as opposed to its mobile AP endeavors, was fundamentally driven by its unwavering concentration on GPUs, the selection of high-growth applications where those strengths could be most effectively leveraged, and the establishment of long-term, strategic partnerships.

One aspect that merits special emphasis is the concept of **'Time'**. In the early 2010s, NVIDIA experienced a challenging period marked by intensified competition, quality issues, and the fallout from the Tegra debacle. Many companies, when confronted with such adversity, tend to prioritize immediate profit recovery, often at the expense of long-term innovation. In contrast,

NVIDIA maintained a long-term vision, investing consistently in deep learning and advanced GPU technology despite short-term setbacks. This willingness to **endure a lengthy, "time-consuming process"** ultimately proved to be the decisive factor behind its remarkable turnaround and subsequent market leadership.

The notion that **"Time coupled with patience is an asset"** is critical in semiconductor ecosystem dynamics, where product development cycles can span several years and rapid technological shifts are the norm. NVIDIA's experience underscores that sustainable competitive advantage is not achieved through short-term measures but rather through persistent, long-term investments in innovation. In an industry characterized by rapid obsolescence and fierce competition, the ability to persevere and continuously adapt over time—while maintaining a clear strategic vision—can differentiate a market leader from its competitors.

Reflective Considerations

What is your company's or your personal core strength within the semiconductor landscape? How can you integrate the lessons learned from NVIDIA (particularly the importance of focusing on core competencies and maintaining a long-term strategic vision) into your own business model and market strategies?

Model 4.
Set Company-Led Strategy in Technological Commoditization

1) Business Model Analysis through The Extended 5 Forces

In this model, we examine how A Co., a tablet PC set maker, bypasses traditional intermediaries (specifically module companies) to engage directly with semiconductor suppliers. This direct negotiation process is analyzed within the extended 5 forces framework, which enables a comprehensive assessment of the dynamics between set companies and semiconductor component suppliers. Moreover, we explore the conditions under which set companies can adjust their relationships with component suppliers to optimize their overall cost structure and supply stability.

A notable characteristic of leading set companies, such as A Co., is their tendency to establish close, collaborative relationships with component manufacturers during the early phases of product development. At this stage, the components in question are typically highly complex and require significant investment in terms of technology and capital. Consequently, the primary objective for these set companies is to achieve successful mass production under stringent quality and reliability standards. However, as products mature and new models are introduced, the underlying technology may gradually transition into a commoditized phase.

At this juncture, set companies must decide whether to vertically integrate—by developing and managing the production of key components in-house—or to diversify their existing semiconductor and module suppliers in order to drive cost reductions through competitive pricing.

When set companies opt for vendor diversification as a means

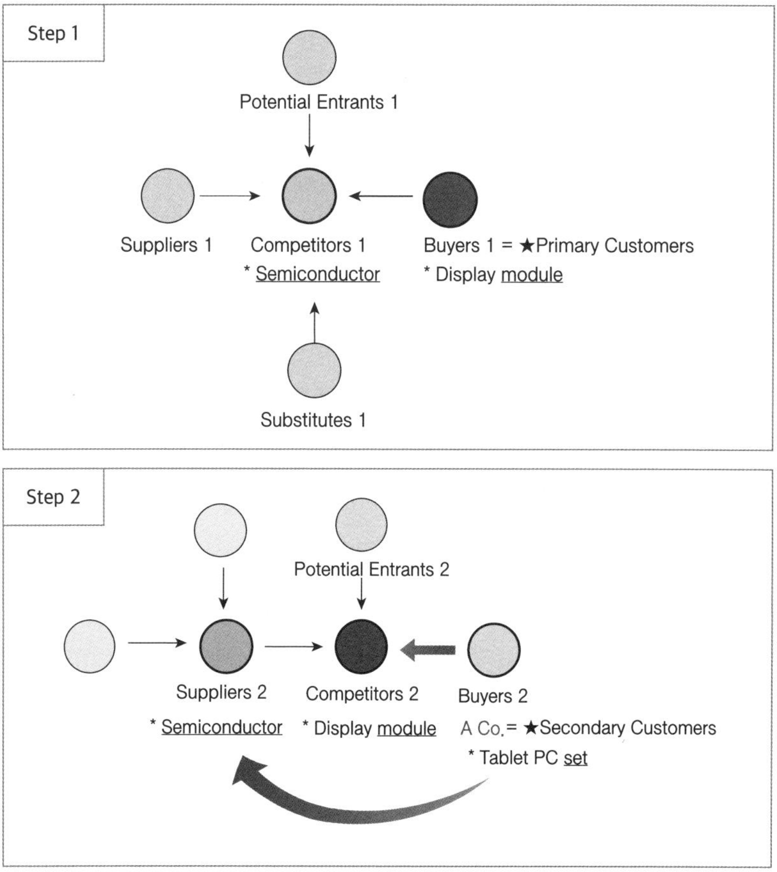

Figure 7 : Extended 5 forces framework for model 4

to reduce costs via price competition, they face both advantages and risks. On the one hand, diversification encourages competition among component manufacturers, thereby exerting downward pressure on prices and strengthening the set company's bargaining position. On the other hand, introducing new suppliers may carry the risk of compromising component quality or disrupting supply stability if the new vendors are unable to meet rigorous performance or yield standards.

Figure 7 depicts the extended 5 forces framework as applied to Model 4. In **Step 1**, the analysis focuses on display module companies acting as Buyers 1, with semiconductor companies operating as Competitors 1. At this initial level, the influence of primary customers on semiconductor suppliers is evident; however, the competitive structures remain somewhat opaque. Hence, an analysis using the extended 5 forces in Step 2 becomes essential.

In **Step 2**, the competitive landscape expands significantly with the inclusion of the set company, A Co., as a secondary customer (Buyers 2). Here, display module companies are reclassified as Competitors 2, and semiconductor companies assume the role of Suppliers 2. A Co. deliberately bypasses the module companies in order to negotiate directly with semiconductor suppliers. This strategic move is designed to secure greater control over the supply chain and to enhance price negotiation leverage.

In practice, **A Co.** was able to exercise what can be termed **"strategic leverage"** reshaping supplier relationships to better suit its cost and production requirements.

Vendor diversification, both among semiconductor companies and module integrators, is a common and natural strategic maneuver among set companies. Nevertheless, when a set company embarks on diversifying its component suppliers, it must carefully evaluate a range of factors. These include the technological competitiveness of the prospective partners, their production capacities, execution capabilities in business, and the nature of the relationships between these suppliers and competing set companies. Conversely, semiconductor companies need to anticipate potential shifts initiated by secondary customers and prepare differentiated strategies in advance to proactively respond to changes imposed by both set companies and module manufacturers.

At the core of this analysis is the need to understand the conditions under which set companies are capable of directly controlling key components. Typically, set companies move to directly control components when the underlying technology has reached a state of commoditization—that is, when other semiconductor competitors are able to produce similar performance products at scale, thereby significantly increasing competitive pressure. For semiconductor companies, analyzing this commoditization process has traditionally involved assessing not only whether competitors are already mass-producing products but also precisely estimating the timing for when new competitors might launch their samples during the development stage and ultimately transition to mass production.

However, recognizing that traditional analysis has often neglected the quantification of technology commoditization's impact on profit and revenue, **I propose the concept of Technology Commoditization Speed.** This metric reflects the rate at which a specific technology becomes widely adopted in the market, undergoes price reductions, and experiences shifts in market share. It is defined as,

Technology Commoditization Speed
= Price Reduction Rate / Market Share * 100

- ✓ **Price Reduction Rate:** As technology becomes commoditized, the number of companies offering the same solution increases, intensifying competition and accelerating price reductions. A higher rate signifies rapid commoditization.
- ✓ **Market Share:** With increased commoditization, more competitors enter mass production using similar technology, which typically results in a decline in an individual company's market share.

This ratio serves as a simple yet effective indicator of how quickly technology is transitioning from a differentiated asset to a commodity.

It is recommended for companies with a market share of at least 20%, since lower shares may yield disproportionately high values that can distort analysis. For companies yet to reach this threshold, the analysis should focus on competitors with established, higher market shares. From the perspective of both primary and secondary customers, the strategic imperative often centers on reducing the market share of the leading component supplier. Therefore, closely monitoring trends among these leading suppliers is essential.

Table 8 : Case Analysis

Case 1	2022	2023	2024	2025	Average
Price Reduction Rate	2%	5%	6%	6%	5%
Market Share	65%	60%	55%	50%	58%
Technological Commoditization Speed	3.1	8.3	10.9	12.0	8.6

Case 2	2022	2023	2024	2025	Average
Price Reduction Rate	2%	2%	2%	2%	2%
Market Share	30%	32%	32%	35%	32%
Technological Commoditization Speed	6.7	6.3	6.3	5.7	6.2

Furthermore, **Table 8** provides a detailed case analysis. In Case 1, the average technology commoditization speed—calculated based on data from 2022 to 2025—was found to be 8.6, with a noticeable annual increase. This trend indicates that the leading component company, which initially held the highest market share, faced escalating pressure from customers to reduce prices, thereby accelerating commoditization. By 2025, the speed had increased to 12, suggesting that competitive entry was intensifying. In contrast, Case 2 exhibited a stable price reduction rate with a slight increase in market share, indicative of a company with stronger competitive positioning that could maintain or even expand its market share despite price pressures.

Table 9 : Comparison of technology commoditization speed

Category	Case 1	Case 2
Price Reduction Rate	High (continuously increasing)	Same
Market Share	Decreasing (customer pressure)	Slight increase
Technological Commoditization Speed	Fast (8.6 ~ 12)	Slow (stable)
Market Share Change	Continuous decrease	Slight increase
Conclusion	**Technology commoditization accelerated due to intensified competition**	Stable technology commoditization speed due to strong competitiveness

Table 9 summarizes these comparisons, illustrating the differential impacts of commoditization speed on strategic positioning and profitability. As technology commoditization accelerates, semiconductor companies find it increasingly challenging to differentiate their products, as competitors' technological, manufacturing, and operational capabilities converge. Ultimately,

this convergence leads to **heightened price pressures** from both secondary (set companies) and primary customers. Therefore, semiconductor companies must critically assess how far they are willing to respond to these pressures and adjust their long-term business strategies accordingly.

To address these challenges, a strategic pricing tool called the **PLC-P graph** (Product Life Cycle-Profit) is introduced. This tool combines an analysis of a **product's life cycle**—from research and development (R&D) to introduction, growth, maturity, and eventual decline—with its associated **profit** trajectory.

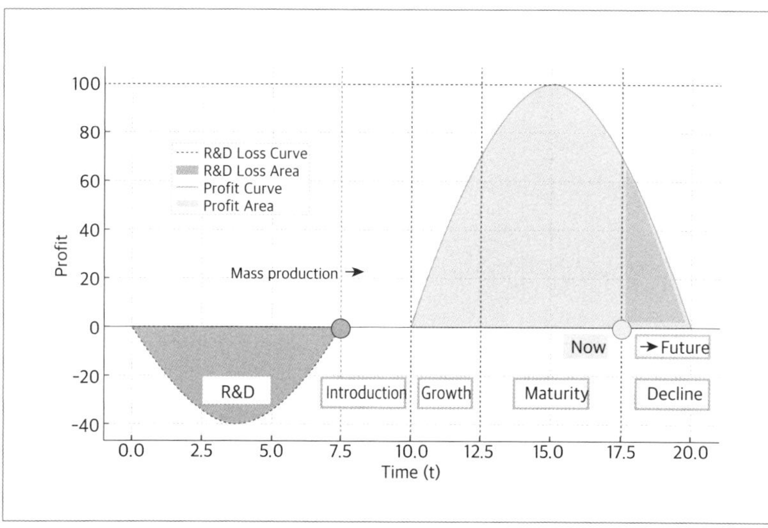

Figure 8 : PLC-P graph (Product Life Cycle with Profit)

Figure 8 illustrates a generalized PLC-P graph derived from semiconductor industry experience. In this representation, the Y-axis denotes operating profit and the X-axis represents time. The graph delineates the five distinct stages of the product life cycle: R&D, Introduction, Growth, Maturity, and Decline. Notably, the R&D phase incurs significant losses, while the growth stage sees a sharp increase in profit, followed by a plateau and eventual decline during maturity. For instance, in the presented example, losses from t=0 to t=7.5 amount to -200, and once mass production commences, the operating profit rises to +500—yielding an overall net effect of +300. This model helps companies assess how price adjustments affect profitability throughout the product life cycle.

By graphically representing these dynamics, companies can develop more systematic pricing strategies that integrate both short-term adjustments and long-term planning. For example, a company that has already achieved substantial profits during the growth phase might adopt an aggressive pricing strategy to defend its market share during maturity. Conversely, a latecomer competing with an established leader may need to maintain higher prices to recoup earlier investments, even as it negotiates with set makers.

To synthesize the insights, key factors in **price decision-making using the PLC-P graph** include,

a) **Comparing** the targeted profit with the expected profit for the product under a proposed price reduction.
b) **Simulating** the PLC-P trajectory for upcoming new products, extending beyond current mass production.
c) **Developing** a comprehensive strategy that integrates these insights to minimize competitor entry and optimally balance market share and profit.

Given that companies routinely receive price reduction requests, it is crucial that they do not focus solely on short-term profit requirements while ignoring the historical performance and future potential of a product. The PLC-P graph thus becomes an indispensable tool for aligning immediate pricing decisions with long-term strategic objectives.

2) Key Force and Market Diversification Strategies

Table 10 summarizes the extended 5 forces and market diversification factors relevant to this model. Analysis through the extended 5 forces framework reveals that the key competitive element in this scenario is the **bargaining power of secondary customers** (Buyers 2). These customers exert direct influence over semiconductor companies by dictating supply conditions and price competitiveness. Accordingly, this business model is named **"Set Company-Led Strategy in Technological Commoditization."**

Table 10 : Extended 5 forces and market diversification factors for model 4

<table>
<tr><th>Component</th><th colspan="2">Factor</th><th>Business Model</th></tr>
<tr><td rowspan="2">Extended 5 Forces Theory</td><td>Key Force</td><td>Bargaining power of secondary customers (Buyers 2)</td><td>◎</td></tr>
<tr><td colspan="2">Strength of the Competitive Forces for Semiconductor</td><td>◎</td></tr>
<tr><td rowspan="6">Suppliers 2 analysis (Semiconductors)</td><td rowspan="3">Technological innovation</td><td>Technological innovation</td><td>○</td></tr>
<tr><td>Successful transition from research to development</td><td>△</td></tr>
<tr><td>Enhancing yield</td><td>△</td></tr>
<tr><td rowspan="3">Diversification</td><td>Customer</td><td>○</td></tr>
<tr><td>Application</td><td>○</td></tr>
<tr><td>Product</td><td>◎</td></tr>
</table>

When the commoditization speed of technology accelerates rapidly, semiconductor companies face intensified competition as price pressures increase. In such a scenario, companies must anticipate the influence of set makers on component pricing and strategically deploy tools such as the PLC-P graph to craft effective pricing proposals. These strategies help in maintaining differentiated competitiveness even as technology becomes commoditized and market dynamics shift.

Moreover, it is imperative for semiconductor companies to

continuously evaluate the commoditization speed of the technologies or products they develop. If this metric indicates a rapid transition, firms should employ the PLC-P graph to forecast future market conditions and adapt their strategic responses accordingly. This proactive approach ensures that companies are not only defending against short-term price cuts but are also positioning themselves to exploit long-term market opportunities.

In conclusion, **the extended 5 forces framework** combined with the analysis of **technology commoditization speed** and the strategic application of the **PLC-P graph** provides a comprehensive method for understanding and responding to the challenges posed by technological commoditization in the semiconductor industry. This integrated strategy empowers semiconductor companies to balance the pressures from secondary customers while sustaining long-term competitive advantages.

Model 5.
Strategic Alignment for Commercializing New Technology

1) Business Model Analysis through The Extended 5 Forces

In 2017, to commemorate the 10th anniversary of its mobile phone launch and to achieve product differentiation, A Co., a mobile phone set maker, embarked on a transformative strategy to transition its display technology from LCD to OLED. During this period, OLED display modules were supplied by SD Co., while the corresponding semiconductor components were developed through a strategic partnership with S Co. This collaborative arrangement exemplifies the power of aligning technology transitions with strategic supplier partnerships.

Figure 9 illustrates the extended 5 forces framework as applied to this model. At the initial analytical stage (**Step 1**), the focus is on the direct relationship between S Co. and SD Co.—the analysis reveals that the primary competitive forces are limited to the immediate supplier–customer interaction. In this conventional view, the semiconductor supplier, S Co., and the OLED module provider, SD Co., are engaged in a relationship that appears to revolve solely around product promotion and order acquisition. However, such a perspective is inherently narrow.

By advancing to **Step 2** of the extended 5 forces analysis, we incorporate a critical additional dimension: the influence of

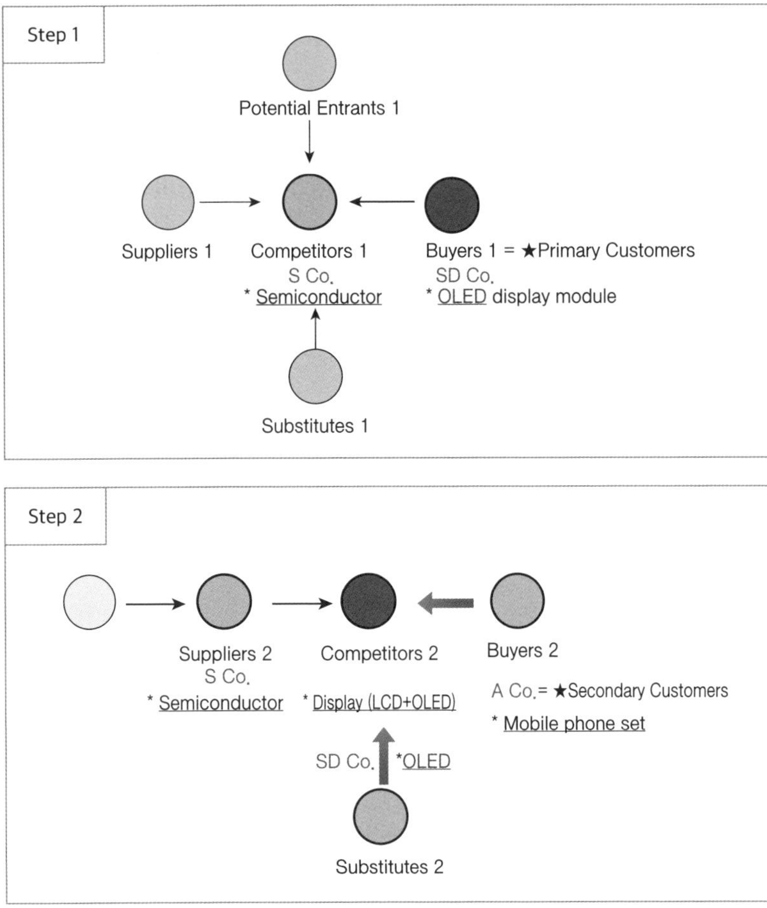

Figure 9 : Extended 5 forces framework for model 5

secondary customers. In this context, Buyers 2 (A Co.)—which had previously depended exclusively on LCD display technology—began to adopt OLED technology, supplied by SD Co., as a newly emerging Substitute 2. As a result, SD Co. eventually becomes classified as Competitor 2. Consequently, the competitive landscape expands, with Competitors 2 now comprising both the incumbent LCD technology and the emerging OLED solution. As a result, S Co., initially a Competitor 1 in Step 1, is reclassified as a Supplier 2 in Step 2, as SD Co. becomes Competitor 2 by entering A Co.'s display supply chain.

The introduction of new technology in the form of OLED as a substitute disrupts the conventional allocation of demand for display modules. The strategy pursued by A Co.—as a secondary customer—directly influences the competitive dynamics, with consumer end-preferences determining whether OLED or LCD models will dominate the market. As a result, **forecasting set demand becomes increasingly complex**, particularly when new models have no historical performance data. **Set companies**, when differentiating new models that incorporate breakthrough technologies, **typically set ambitious sales targets**. If semiconductor companies scale production based on these optimistic projections, they risk facing significant supply–demand mismatches.

Moreover, one of the most critical challenges when launching new products that integrate novel technologies is managing the **yield rate**. In the case of OLED displays, the yield is both paramount and highly variable. Should the actual yield of a customer's module fall below the expected threshold, a shortage of semiconductor components may occur, thereby disrupting production continuity. Conversely, yields exceeding expectations may lead to an oversupply of semiconductors, resulting in excess inventory and associated financial pressures. Consequently, semiconductor suppliers like S Co. must develop robust forecasting models to accurately predict the yield performance of their primary customers (module companies) and adjust their production strategies accordingly.

In summary, **the convergence of set demand variability from secondary customers and yield variability from primary customers substantially complicates both demand forecasting and production planning**. To address these challenges effectively, additional response methods and strategic initiatives are discussed in detail in Section 3 of this model.

2) Key Force and Market Diversification Strategies

The strategic decision by the set company to launch a new model incorporating OLED technology introduces a significant disruptive force: OLED emerges as a substitute threat to traditional LCD display technologies. At the same time, it creates new market opportunities for OLED-based solutions. According to the extended 5 forces framework, this evolution underscores how the emergence of a substitute technology can simultaneously act as a threat and as an entry opportunity. This duality forms the core competitive force in this model, as further illustrated in **Table 11**.

Table 11 : Extended 5 forces and market diversification factors for model 5

Component	Factor		Business Model
Extended 5 Forces Theory	Key Force	Threat (Opportunity) of substitute products or services	◎
	Strength of the Competitive Forces for Semiconductor		◎
Competitor 2 analysis (OLED Display)	**Technological innovation**	Technological innovation	○
		Successful transition from research to development	◎
		Enhancing yield	◎
	Diversification	Customer	◎
		Application	△
		Product	○

Table 11 delineates the extended 5 forces and market diversification factors for Model 5. Based on this analysis, the business model is formally named **"Strategic Alignment for Commercializing New Technology."** This designation underscores the critical considerations that semiconductor companies must address when introducing new technologies for set launches. In particular, the model emphasizes the importance of carefully evaluating both the variability in demand for new set models and the associated yield forecasts at the levels of the primary customer (OLED display modules) and the semiconductor supplier (S Co.).

A notable aspect of this model is its forward-looking perspective: for example, **A Co. is planning to launch a foldable phone model in the near future**—an undertaking that will demand a significantly higher level of technical expertise as well as precise volume forecasting compared to the earlier transition from LCD to OLED. By applying the insights derived from this model, semiconductor companies can more effectively tailor their strategies to respond to such new market opportunities.

3) Strategic Considerations Applying New Technology

When semiconductor companies introduce new products that integrate advanced technologies to serve set companies, they encounter a host of interrelated challenges. To achieve successful market entry, firms must adopt multi-faceted strategies that encompass **market and demand forecasting, yield management, and supply chain optimization**. These strategies are not isolated, rather, they are interconnected and must be executed in tandem to ensure overall success.

Market and Demand Forecasting

✓ **Understanding Customer Requirements:** It is essential for semiconductor companies to comprehensively grasp the **technical requirements**, **production schedules**, and **demand patterns** of their customers. Given that the customer base may extend beyond primary customers to include secondary customers, employing the extended 5 forces framework can help capture these nuances.

- ✓ **Analysis of New Set Models**: New technologies are often introduced first in the premium segments of set companies. It is critical to evaluate the anticipated market share and competitiveness of these new models by comparing their performance and pricing against competing products. Demand should be simulated using various scenarios to account for different market responses.
- ✓ **Supply Chain Analysis**: An in-depth analysis of the supply chain—spanning set companies, module manufacturers, and semiconductor companies—is crucial to predict **yield** variability, turnaround times (**TAT**), and **inventory** management requirements. This analysis must be dynamic, with information continually updated as production timelines and yield rates evolve.
- ✓ **Scenario Planning**: By developing multiple demand forecasting scenarios, companies can simulate diverse market situations that may arise during the launch phase of new set models.
- ✓ **Supply Planning and Financial Analysis**: Integrating demand forecasting data with supply planning allows companies to estimate semiconductor volumes, forecast revenue, and determine optimal resource allocation.

Semiconductor Yield Prediction and Management

- ✓ Firms must establish robust management measures to mitigate yield variability during the initial introduction of new technology. This involves significant technical investments and the implementation of comprehensive monitoring systems to stabilize yield performance.

Production Investment and Supply Chain Optimization

- ✓ **Investment Optimization:** Given the high production and capital costs associated with new technology, it is critical to develop strategies that optimize investment and ensure long-term operational stability.
- ✓ **Flexible Supply Chain Management:** Creating a flexible supply chain capable of addressing initial yield fluctuations and quality variabilities enables rapid responses to unforeseen circumstances.

Semiconductor Inventory Management

- ✓ Companies must manage risks associated with inventory surpluses or shortages meticulously, ensuring that fluctuations in customer demand and yield do not disrupt production or lead to significant financial losses.

Collectively, these processes provide a comprehensive framework for semiconductor companies to manage potential variabilities when launching new set models with advanced technology.

Thus far, we have examined strategic considerations in applying new technology. In the following section, these considerations are revisited through the lens of the semiconductor commercialization process, with particular emphasis on the transitional phases that require close and proactive management for robust business execution. Accordingly, the 5:3:2 Rule is proposed as a structured and pivotal framework to support effective management of these transitional dynamics.

5:3:2 Rule

To explore in depth the strategic considerations applying new technology, it is insightful to view them through the lens of **the commercialization process**, which is typically divided into three key stages: Research, Development, and Production. In light of these facets, **I propose the 5:3:2 Rule,** which captures the idea that **strong management of transitional phases (technical hand off) is more impactful** than executing each stage in isolation—particularly in terms of **cost savings. (Figure 10)**

In a simplified view:

✓ **50%:** The transition from research to development is crucial because it **determines product feasibility, design correctness, and process alignment**. For example, if a novel transistor structure or memory cell is proven in the lab but cannot be seamlessly adapted to existing manufacturing lines, the project languishes.

- ✓ **30%:** The progression from development to mass production tests yield ramp and cost optimization. Even if a product's design is functionally validated, ensuring adequate volumes at competitive cost is make-or-break.
- ✓ **20%:** Finally, during mass production, incremental yield improvements or minor cost reductions may matter, but the overarching profitability or timeline has mostly been set by prior phases.

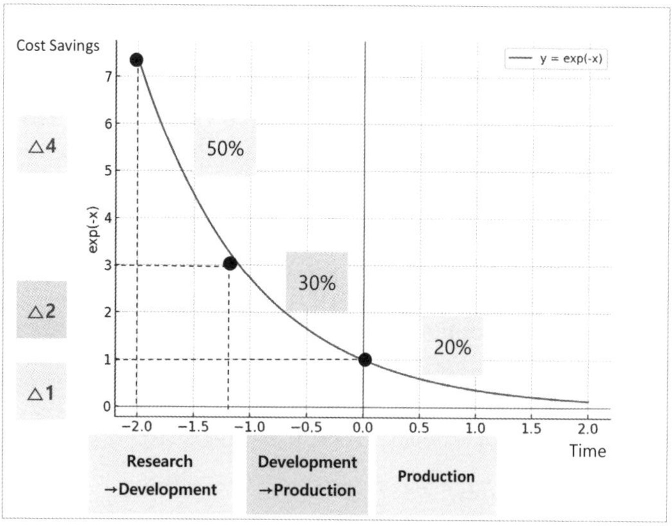

Figure 10 : Cost Savings throughout The Commercialization Process

In real-world operations, advanced logic or memory companies each have slightly different **R&D pipeline structures**. For instance, some might allot extended time for fundamental materials research (like new gate dielectrics or spintronics) before deciding feasibility. Others carry an "agile" approach, integrating advanced research with early design explorations or pilot lines. The ultimate aim, however, remains consistent: bridging each threshold with minimal friction and maximum yield learning.

Commercialization success also hinges upon forging partnerships with EDA vendors, OSAT providers, or top-tier device OEMs early on. If the supplier invests heavily in next-gen technology but lacks anchor customers or robust packaging solutions, the final mass production stage might stall. By carefully mapping these **cross-functional dependencies within the semiconductor ecosystem** and ensuring they are resolved prior to large-scale ramp, a semiconductor firm can expedite time to profitable production.

Because semiconductors require synergy across R&D labs, pilot lines, large-scale fabs, and robust packaging solutions, diligently planning and executing these transition points fosters faster time to market, lower re-spin rates, and an earlier break-even. More importantly, cost savings and margin improvements are frequently determined by how effectively cross-functional teams manage the technology hand off.

Recalling the 5:3:2 Rule: more than half of a product's ultimate

competitive advantage might be locked in during the earliest transitions, giving yield the potential to define an entire product's fate.

In many respects, **yield** is the lifeblood of the semiconductor business. Achieving robust yields within a few quarters after ramp sets the difference between a cost-effective product and one that flounders or fails. For example, a DRAM maker might aim for yields above 90% once the node matures, but if the yield lingers at 60–70% for too long, per-bit costs spike, crippling competitiveness. The same logic holds for specialized logic chips: if an advanced CPU or GPU cannot meet yield targets, the supplier either has to raise prices (risking lost design wins) or absorb margin losses.

Yield likewise influences reliability: poorly optimized processes can result not only in immediate test failures but also higher field returns or unexpected performance degradation. Over time, such fiascos can tarnish brand reputation and sabotage future design sockets. Conversely, a proven track record of stable yields builds trust with top-tier OEMs or HPC platform developers, paving the way to deeper co-development or multi-product deals.

Hence, in any semiconductor program, yield management must be integrated with the broader strategic approach—coordinating R&D, pilot-line expansions, test engineering, and supply chain scheduling.

Model 6.
Strategic On-Site Dispatching for Customer Diversification

1) Business Model Analysis through The Extended 5 Forces

As technology commoditization accelerates in a specific semiconductor industry, entry barriers lower, and competition among semiconductor companies intensifies. Consequently, companies should pursue customer diversification to secure revenue and profits. However, in an environment of heightened competition, diversifying customers becomes a significant challenge for companies. In this business model, we analyze a case study of S Co.'s attempt to diversify its customer portfolio by supplying semiconductors for SH Co.'s TV panels.

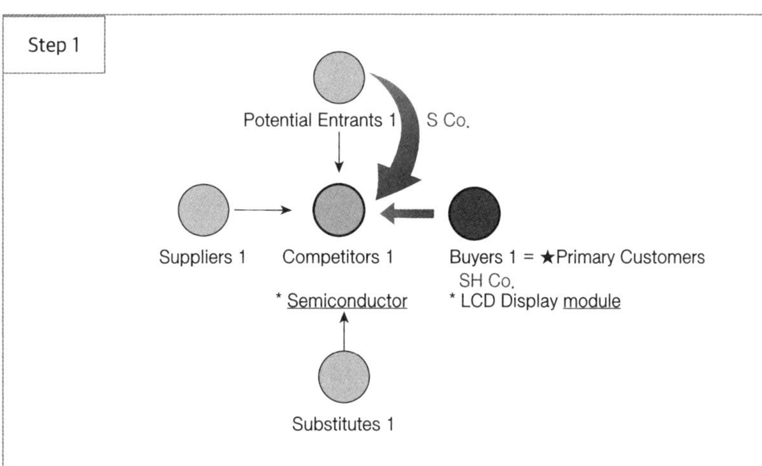

Figure 11 : Extended 5 forces framework for model 6

Figure 11 depicts the extended 5 forces framework applied to this case. Compared to other models, this one has a relatively simple structure, with S Co. attempting customer diversification by entering SH Co.'s supply chain. By incorporating this process, S Co. repositions itself within the market—from merely being a potential entrant to becoming an active competitor within SH Co.'s supply ecosystem.

Around 2010, the semiconductor market for TV panels began experiencing accelerated commoditization. At this juncture, differentiating one's technology became increasingly challenging, forcing companies to explore innovative strategies for customer diversification. Recognizing that its traditional customer base was offering limited growth opportunities, S Co. identified SH Co.—a major display panel manufacturer with a market share of roughly 20–30%—as a strategic target despite having no prior relationship. Even though SH Co. initially held a somewhat unfavorable view of S Co., the potential for diversification was too significant to ignore.

To penetrate this new market, **S Co. implemented a customer cultivation strategy characterized by long-term on-site deployment.** This strategy involved dispatching dedicated teams of technical marketing experts, sales personnel, and development engineers to establish a persistent, physical presence at SH Co.'s overseas factories and offices. The primary goals of this deployment were to build trust, actively gather insights about customer requirements, and secure a direct channel into mass production.

Field personnel engaged in a variety of tactics to establish rapport. They participated in informal settings—such as cafeterias, break rooms, and even waiting under the customer's entrance awnings during inclement weather—to seize every opportunity for a personal interaction. In many instances, they waited for extended periods (sometimes 5, 10, or even 12 hours) to meet with key decision-makers. This intensive engagement was not merely a matter of persistence; it served as a critical bridge, fostering mutual understanding and enabling real-time exchange of strategic and technical information between S Co. and SH Co.

Furthermore, the on-site teams maintained continuous communication with S Co.'s headquarters, ensuring that insights gathered from the field were promptly integrated into strategic planning and product development. This rigorous approach helped S Co. refine its customer diversification strategy, enabling it to respond dynamically to evolving customer needs and market trends. It also highlighted the inadequacy of relying solely on publicly available information, which often lacks the nuanced context necessary for effective decision-making.

Practical Guide for On-Site Deployment Activities

Given the complexity of customer diversification in a commoditized market, S Co. developed a detailed, multi-step framework for on-site deployment. This framework is designed to maximize the impact of field activities and is structured as follows,

Step 1. Project Organization and Planning

- ✓ **Project Naming:** Establishing **a strategic project name motivates** internal teams, reinforces a shared sense of purpose, and aligns both short-term execution and long-term strategic vision.
- ✓ **Objective Definition:** Establish clear, quantifiable objectives tailored for successful entry into new customer projects, with the ultimate goal of securing mass production.
- ✓ **Operational Road Map:** Develop a comprehensive plan that outlines key milestones, deadlines, and the allocation of resources, including cross-departmental collaboration from engineering, sales, and supply chain teams.
- ✓ **Resource Allocation**: Identify the necessary resources—including personnel, technology, and capital—and assign responsibilities to ensure that every facet of the deployment is covered.

Step 2. Customer Organizational Analysis

- ✓ **Mapping the Organizational Structure:** Conduct a thorough analysis of the customer's organizational chart to understand **decision-making processes and hierarchies,** while also **pin-**

pointing key influencers.

- ✓ **Identification of Key Stakeholders:** Identify individuals who hold significant influence over purchasing decisions (Key Persons), allies who are supportive (Knights), and potential detractors (Antis). This knowledge enables S Co. to tailor its engagement strategy effectively.
- ✓ **Cultural and Strategic Assessment:** Analyze the customer's corporate culture and strategic priorities to align S Co.'s proposals with the customer's long-term objectives.

Step 3. Strategy Implementation and Field Operations

- ✓ **On-Site Presence**: Establish a permanent or semi-permanent base near the customer's facilities, such as an office or hotel, to facilitate frequent and informal interactions.
- ✓ **Information Sharing**: Set up robust communication channels between on-site teams and headquarters to ensure that real-time insights are captured, analyzed, and acted upon promptly **in order to secure mass production**.
- ✓ **Feedback Mechanisms**: Implement regular reporting structures and feedback loops to monitor progress, address emerging issues, and adapt the strategy as needed.

Step 4. Evaluation and Adaptation

- ✓ **Data-Driven Analysis**: Continuously assess customer interactions using both **qualitative and quantitative metrics**.
- ✓ **Strategic Flexibility**: Remain agile and willing to pivot the strategy based on changing market conditions or new information regarding customer requirements.
- ✓ **Cross-Department Collaboration**: Encourage collaborative decision-making within S Co., ensuring that all relevant departments contribute to and are informed by the on-site findings.

Table 12 : Practical guide for on-site deployment activities

	Detail 1	Detail 2
Step 1	Project Planning: Objectives, Goals, Operation plan, Participating departments, Strategy, Project name	Information sharing, Reporting, Communication System & Action items
Step 2	Understanding customer organization and Decision making process	- Identifying customer key man, knight, anti - Analyzing the structure of the customer's development, purchasing, and executive departments by model
Step 3	Implementing strategy for mass production project acquisition	Executing with the goal of securing mass production orders
Step 4	Evaluating customer information by collective intelligence.	Assessing the exclusivity, feasibility, and success potential

Table 12 (referenced in the text) provides a detailed breakdown of these steps, serving as a practical guide to ensure the success of on-site deployment activities. Such a structured approach not only minimizes the risk of miscommunication but also maximizes the chances of converting initial contacts into long-term, mutually beneficial partnerships.

Case Study Analysis: Customer Diversification in Practice

Building on its comprehensive on-site deployment strategy, S Co. eventually received a proposal from SH Co. that offered two distinct strategic options.

- ✓ **Option 1**: Develop a semiconductor for low-spec displays, priced lower than competing products, thereby securing immediate, short-term demand. However, this option was anticipated to experience a decline in demand over the long term due to plans by SH Co. to transition to high-spec products.
- ✓ **Option 2**: Develop a semiconductor for high-spec displays, which, although associated with lower initial demand, promised greater long-term growth potential if S Co. could establish a competitive advantage in both development and pricing.

After extensive deliberations and in-depth discussions between on-site personnel and S Co. headquarters, the company opted for Option 1. This decision was driven by several strategic factors: the lower technical complexity and **more immediate scalability** of low-spec products, as well as the necessity to **secure a foothold in mass production**. S Co. recognized that successful mass production is paramount in the semiconductor industry, and that establishing a reliable, high-volume supply base would lay the foundation for future growth. In addition, S Co. was acutely aware of the risks inherent in entering high-spec components market with existing competitors who already had established mass production processes. In particular, there was the possibility that S Co. might lack sufficient experience in mass-producing high-spec components, which could lead to operational challenges and quality issues.

Option 1 was addressed through the on-site deployment strategy, which enabled S Co. to build trust with SH Co. by demonstrating unwavering commitment—exemplified by field personnel who would wait under the customer's entrance even on rainy days and engage in prolonged informal meetings. This level of dedication ultimately fostered a strong mutual understanding, which proved critical when semiconductor competitors gradually withdrew from SH Co. due to intensifying price competition.

As a result, S Co. became the sole supplier of the low-spec product, providing it with the leverage to raise prices and secure enhanced profit margins. This case vividly illustrates **the importance of strategic on-site deployment as a means to achieve customer diversification**, particularly in environments where the relationships with existing customers are already well established and fiercely competitive.

2) Key Force and Market Diversification Strategies

Table 13 : Extended 5 forces and market diversification factors for model 6

Component	Factor		Business Model
Extended 5 Forces Theory	Key Force	Bargaining power of buyers	◎
	Strength of the Competitive Forces for Semiconductor		◎
Competitors 1 analysis (S Co.)	**Technological innovation**	Technological innovation	△
		Successful transition from research to development	△
		Enhancing yield	△
	Diversification	Customer	◎
		Application	△
		Product	○

Table 13 summarizes the extended 5 forces and market diversification factors for this model. In analyzing the framework, the key competitive element identified is the bargaining power of buyers—specifically, the primary customers who ultimately determine which vendors are selected to supply critical com-ponents. This model emphasizes a customer diversification strategy that is implemented through what we term **"Strategic On-Site Dispatching for Customer Diversification."** In practice, this strategy involves deploying dedicated personnel to the customer's site to establish deeper relationships, gather nuanced insights, and integrate these insights into the overarching business strategy.

In mature markets, the establishment of long-term on-site deployments is indispensable. Such an approach not only facilitates the collection of granular customer intelligence but also enables continuous adaptation of the company's strategic initiatives in response to evolving market conditions. To maximize the efficacy of this approach, it is essential to institute a well-structured support system that spans from top-level executives to field staff. This system ensures that on-site personnel receive the necessary resources, guidance, and support to effectively engage with customers and address their complex requirements.

Special Theme:
Semiconductor Price Negotiation

In the realm of business, particularly within the semiconductor industry, the art of price negotiation represents the pinnacle of strategic execution. Far from being a mere transactional exercise, semiconductor price negotiation can be likened to the creation of a masterpiece. It demands a sophisticated blend of strategic foresight, perseverance, patience, composure, and the ability to craft persuasive narratives. Drawing on both successes and setbacks, this section elucidates key considerations from the perspective of a semiconductor company engaged in complex B2B negotiations.

1. Who Determines Semiconductor Prices in B2B?

At first glance, one might assume that semiconductor prices are determined solely by the direct negotiations between the customer's purchasing or executive teams and the supplier's sales force. However, the reality is far more intricate. Within the customer's organization, multiple internal factors exert influence. The purchasing department, while primarily responsible for managing prices, operates under the significant pressure of several influential forces: the CEO's strategic directives, cost-reduction mandates imposed by the accounting team (or CFO), and technical demands articulated by the development team. Moreover, individual material managers within the purchasing department often find themselves in competition with one another to meet stringent cost-reduction targets. These internal dynamics create a scenario in which the purchasing representative, sitting across from the semiconductor company's marketing or sales team, is under immense pressure to negotiate lower prices.

Concurrently, the semiconductor company's marketing and sales teams are driven by revenue and profit targets, which adds another layer of complexity to the negotiation process. Thus, **the final semiconductor price is the result of a highly complex, multi-dimensional negotiation process that spans various departments and hierarchical levels on both sides**.

2. Strategic Storytelling in Price Negotiations

In addressing price negotiations, semiconductor companies must adopt a strategy that transcends mere number-crunching and instead embraces the art of storytelling. As famously noted by Professor Patrick Winston, **"Humans are storytelling animals."** In the context of price negotiations, this means **constructing a narrative** that weaves together multiple strategic terms and conditions into a coherent, persuasive proposal.

For example, a semiconductor company might internally prepare proposals that include,

✓ A commitment to meet the customer's increased production

volume targets for the current quarter, thereby justifying a moderate price reduction.

- ✓ An agreement to maintain stable prices for the current period in exchange for future price concessions on a new product launch.
- ✓ Negotiating a modest price cut contingent upon a guarantee that the customer's market share will surpass a certain threshold in the upcoming fiscal period.

Such proposals are designed not to dictate terms in a rigid, linear fashion but rather to open a dialogue that allows both parties to explore a range of mutually beneficial options. In practice, **proposals** are often intentionally framed in a way that is somewhat **ambiguous, leaving room for interpretation and further discussion**. This ambiguity encourages the customer's internal debate and reflection, rather than forcing an immediate, overly definitive response. If the customer were to respond with an unequivocal "Let's go with that," it might signal that the proposal was overly generous or that the supplier missed an opportunity to negotiate more favorable terms.

Moreover, **during negotiations, subtle cues such as intonation, choice of words, and body language can provide critical insights into the customer's true position.** It is, therefore, imperative that following any customer meeting, the semiconductor company conducts an internal debrief to collectively interpret these signals and adjust its strategy accordingly.

3. Keep the Final Card to Yourself

Another critical strategic principle in semiconductor price negotiations is to retain a "hidden card" until the final stages of the discussion. At the outset, both the customer and the semiconductor company appear to have rigid, non-negotiable bottom lines. However, as negotiations progress and various strategic terms are introduced, a dynamic environment emerges where concessions are made incrementally. In this context, it is advisable for the negotiator to withhold a final, unexpected proposal or figure until the concluding phase of the negotiation process. This **"hidden card"** is not intended to withhold information from internal teams; rather, it represents a **tactical reserve**. As the negotiator remains **deeply engaged in the negotiation process**, a creative idea may emerge—one that contributes meaningfully to the business and benefits the company.

Ideally, this hidden card should address a broader aspect of the business relationship beyond the immediate product price. By the time it is introduced, both parties will have engaged in extensive discussions covering various components of the deal. The sudden presentation of an innovative term or figure—unexpected by the customer—can serve as a decisive move to secure more favorable conditions

Key Strategies for Semiconductor Price Negotiation

To summarize, the following key strategies are recommended for effective semiconductor price negotiation,

a) **Understand the Multiple Influences on Price Determination:** Recognize that price negotiations are influenced by a constellation of factors within the customer's organization, including input from the CEO, CFO, and development teams.

b) **Frame Proposals as Strategic Narratives:** Develop proposals that are embedded in a compelling strategic narrative, integrating various terms and conditions in a manner that encourages further discussion.

c) **Maintain Strategic Ambiguity:** Craft proposals that are enticing yet deliberately ambiguous, fostering internal debate within the customer's organization.

d) **Employ a Multi-Angle Negotiation Approach:** Approach negotiations from various perspectives, continuously adapting based on subtle cues and real-time feedback.

e) **Retain a Hidden Card for the Final Moment:** Strategically reserve a final, unexpected proposal to be deployed at the closing stages of negotiations, thus leveraging any remaining bargaining power.

This comprehensive approach to semiconductor price negotiation, which integrates detailed market insights and advanced strategic tactics, enables companies to navigate the complex and dynamic B2B landscape effectively.

Model 7.
Customer Relationship Retention via Product Diversification

1) Business Model Analysis through The Extended 5 Forces

Semiconductor companies not only strive to secure large-scale projects that yield significant revenue and profit, but they also recognize that the long-term success of such projects depends on maintaining robust, enduring customer relationships. One critical strategy that emerges in this context is **product diversification**—offering an expanded portfolio of semiconductor products to the same customer. This approach not only drives additional revenue streams but also adds critical strategic value by **reinforcing the customer relationship and providing opportunities for future collaboration**.

For example, in Model 2, S Co. successfully secured a large-scale project by supplying a core semiconductor product (SoC) to G Co. Building on this achievement, S Co. strategically convinced G Co. to adopt additional product lines—specifically Power IC and CIS—as expansion products. Given that G Co. was in the nascent stages of entering the mobile phone market, there was minimal resistance to adopting supplementary products based on the established relationship from the core project. This case illustrates how product diversification within a single customer can create a multipronged revenue channel and reinforce a company's market presence.

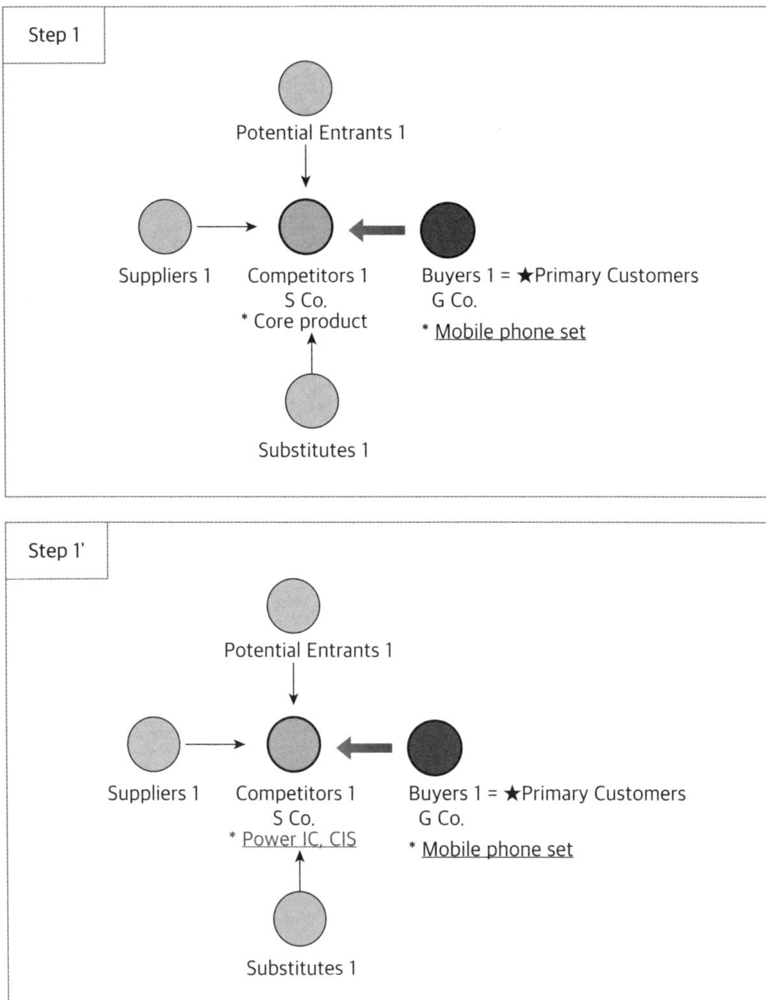

Figure 12 : Extended 5 forces framework for model 7

Figure 12 presents the extended 5 forces framework as applied to this model. In **Step 1** of the analysis, S Co. is depicted as supplying its core product to the primary customer, G Co. (Buyers 1), under a competitive scenario in which S Co. is categorized among Competitors 1. In a subsequent analytical layer, denoted as **Step 1'** in the framework, S Co. extends its product offerings to include additional semiconductor solutions—namely Power IC and CIS—effectively diversifying its portfolio within the same customer. Because this model focuses on a single customer relationship, further subdivision using the extended 5 forces is not required.

As the business environment evolves, S Co. encountered an internal strategic shift at G Co., which led to a reduction in the anticipated market share for the original core product. In such a scenario, having diversified its product offerings, S Co. was able to maintain a robust connection with G Co. through the continued supply of its expansion products. This ongoing relationship enabled S Co. to remain well-informed about G Co.'s future product strategies and road maps, thereby ensuring that the company could adapt quickly to market changes and secure additional business opportunities. In essence, **this model underscores the strategic value of product diversification as a means of retaining customer relationships over the long term.** In an environment where customer demands, competitive landscapes, and market conditions are in constant flux, offering a diversified product portfolio ensures that semiconductor companies remain integrated with their customers and are better positioned to anticipate future opportunities.

2) Key Force and Market Diversification Strategies

Table 14 : Extended 5 forces and market diversification factors for model 7

<table>
<tr><th>Component</th><th colspan="2">Factor</th><th>Business Model</th></tr>
<tr><td rowspan="2">Extended 5 Forces Theory</td><td>Key Force</td><td>Bargaining power of buyers</td><td>◎</td></tr>
<tr><td colspan="2">Strength of the Competitive Forces for Semiconductor</td><td>○</td></tr>
<tr><td rowspan="6">Competitors 1 analysis (S Co.)</td><td rowspan="3">Technological innovation</td><td>Technological innovation</td><td>○</td></tr>
<tr><td>Successful transition from research to development</td><td>○</td></tr>
<tr><td>Enhancing yield</td><td>○</td></tr>
<tr><td rowspan="3">Diversification</td><td>Customer</td><td>△</td></tr>
<tr><td>Application</td><td>△</td></tr>
<tr><td>Product</td><td>◎</td></tr>
</table>

Table 14 encapsulates the extended 5 forces and market diversification factors specific to this model. Through a detailed analysis using the extended 5 forces framework, the primary competitive element identified is the bargaining power of buyers—particularly, the primary customers who ultimately decide whether to adopt additional semiconductor products. From a market diversification perspective, this model centers on a product diversification

strategy termed **"Customer Relationship Retention via Product Diversification."**

This approach emphasizes that, even when a major project is secured, sustaining the relationship requires a proactive strategy. For semiconductor companies, collaborating with strategic or major customers to offer an expanded product suite not only generates additional revenue but also reinforces customer relationship management (CRM). By diversifying product offerings, companies can continue to engage with their customers, remain informed of their evolving needs, and effectively anticipate changes in their product strategies. In this manner, product diversification serves as a robust mechanism to preserve long-term business channels and to open new opportunities for large-scale projects in the future.

Nonetheless, this strategy is not without its risks. The diversification process must be executed with caution, as issues such as quality control or delivery delays associated with newly introduced products could negatively impact the supplier's reputation. Any problems in the execution of expansion products may damage trust and, by extension, the long-term relationship with the customer. Therefore, semiconductor companies must rigorously evaluate the competitiveness and production viability of any new products introduced as part of a diversification strategy.

Model 8.
Balancing Diversification against Customer's Substitute Threats

1) Business Model Analysis through The Extended 5 Forces

In previous models, we have explored the merits of customer, application, and product diversification through the lens of evolved competitive frameworks and market diversification strategies. However, in the realm of business strategy, invaluable lessons are derived not only from success stories but also from **failures**. Model 8 examines a case study of a failed attempt by semiconductor company S Co. to diversify its customer base. This case highlights the potential pitfalls and risk factors that may arise when pursuing customer diversification, particularly in the context of substitute technologies.

S Co. endeavored to broaden its customer base by supplying PDP (Plasma Display Panel) semiconductors to P Co., a prominent manufacturer of PDP. Around 2010, the TV display market was sharply segmented into two competing camps: LCD and Plasma Display. The LCD segment was dominated by major players such as Samsung Display, LG Display, and Sharp, while the Plasma segment was led by companies including SD Co. and P Co. S Co. attempted to diversify by targeting P Co. with PDP semiconductors that it was already mass-producing for SD Co.

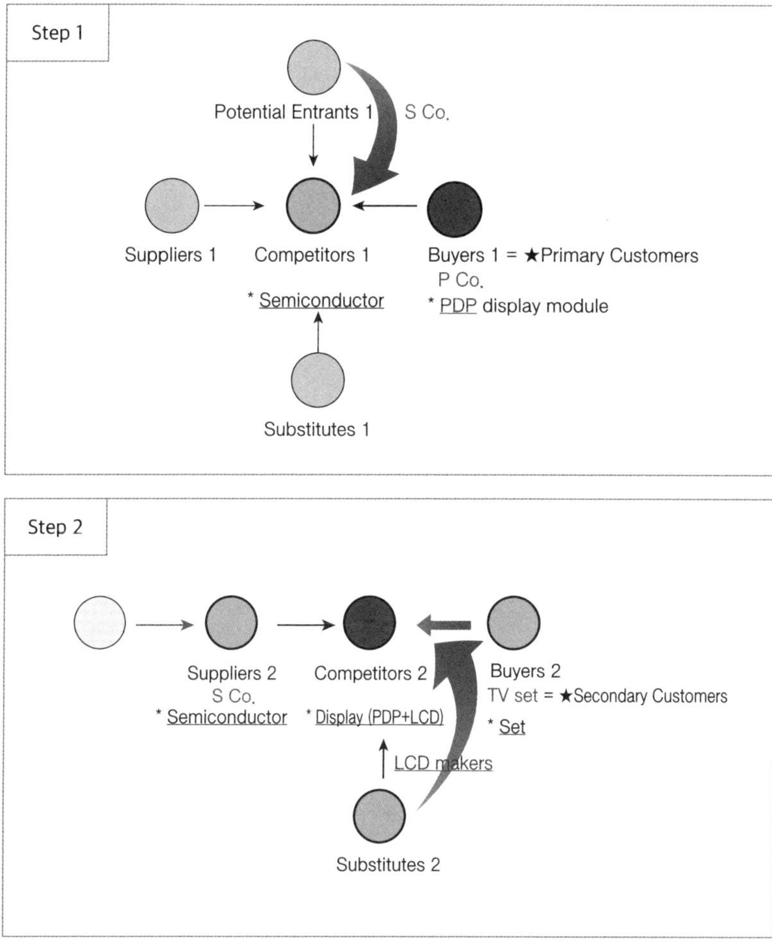

Figure 13 : Extended 5 forces framework for model 8

Figure 13 illustrates the extended 5 forces framework as applied to this model. In the initial stage (**Step 1**), the analysis focuses on S Co.'s direct customer diversification effort—supplying PDP semiconductors to P Co. In the subsequent stage (**Step 2**), the analysis extends to incorporate the influence of secondary customers, specifically TV set companies (Buyers 2), and reclassifies the display suppliers into Competitors 2. From the perspective of PDP manufacturers, LCD technology emerges as a substitute product, thereby complicating the competitive environment. This additional layer of analysis unveils complexities that remain obscured in a conventional five forces analysis, thereby **emphasizing the importance of assessing substitute threats from the standpoint of the primary customer's customer**.

During this period, both the LCD and PDP camps engaged in fierce competition, each advocating the superiority of its own technology and performance. LCD companies forecasted significant increases in LCD demand, while PDP companies anticipated robust growth in PDP demand. However, the overall market demand for TV sets was relatively constrained. Consequently, the combined projected demand from both camps far exceeded the actual market size, leading to **a significant misalignment between expectation and reality**.

Despite the challenging market conditions, S Co. pursued its strategy relentlessly. After an arduous 18-month negotiation process, S Co. secured a new project with P Co. and proceeded with development. Unfortunately, during the initial stages of mass production, S Co. encountered unexpected challenges. Quality issues emerged on P Co.'s module line, and it became evident that resolving these issues would require an extended period of intensive technical intervention. Although S Co. had years of experience in mass-producing semiconductors for other PDP customers and was confident in its product, it faced significant obstacles in addressing the multifaceted quality issues—issues such as static electricity effects—which made it difficult to assign clear responsibility either to the semiconductor or to the customer's module line. Given the Japanese customer's uncompromising emphasis on quality, proceeding with mass production without a definitive resolution was unacceptable.

Faced with these challenges, S Co. was confronted with a stark decision: either fully resolve the quality issues—an endeavor that would necessitate long-term analysis, on-site engineering intervention, and an undefined resolution timeline—or abandon the project entirely, thereby forfeiting the two-year investment of time and resources. Ultimately, S Co. chose to withdraw from the PDP semiconductor business with P Co., accepting a 50:50 cost-sharing arrangement for the defective volume. This decision was driven by the inability to predict the time required to resolve the quality issues, the unsustainability of indefinitely allocating engineering resources, and the risk that continued quality problems would preclude the development of new products.

Coincidentally, during this period, a significant market shift occurred. The PDP camp, which had long competed with LCD, began to suffer considerable losses as LCD technology proved to be superior in terms of performance, cost efficiency, and production volume. Eventually, **P Co. decided to exit the PDP business entirely and closed its PDP factories**. From S Co.'s perspective, in hindsight, failure to achieve mass production might have prevented further complications—such as raw material and IC inventory disposal issues or compensation liabilities—entering the PDP market under these circumstances would have exposed the company to substantial financial and operational risks.

This case study of S Co. underscores several critical insights. First, it is imperative to accurately predict customer demand by not solely relying on the customer's stated requirements, but also **by thoroughly analyzing the competitive dynamics among the customer's own competitors**. Second, customer diversification strategies must incorporate a rigorous assessment of quality risks inherent in a customer's production line, as each customer operates within a unique set of production and evaluation parameters that are often not fully disclosed. Finally, when quality issues arise that demand extensive time and resource commitments to resolve, it may be strategically prudent to withdraw from the project rather than risk further losses.

2) Key Force and Market Diversification Strategies

Table 15 : Extended 5 forces and market diversification factors for model 8

Component	Factor		Business Model
Extended 5 Forces Theory	Key Force	Threat of substitute against the primary customers	◎
	Strength of the Competitive Forces for Semiconductor		◎
Suppliers 2 analysis (S Co.)	**Technological innovation**	Technological innovation	△
		Successful transition from research to development	○
		Enhancing yield	◎
	Diversification	Customer	◎
		Application	△
		Product	○

Table 15 encapsulates the extended 5 forces and market diversification factors pertinent to this model. Through the extended 5 forces analysis, the primary competitive element identified is the threat of substitutes that primary customers (PDP manufacturers) face due to the aggressive competition from LCD technology. This substitute threat, coupled with the inherent risks in customer diversification, necessitates a cautious approach. I propose naming this model **"Balancing Diversification against Customer's Substitute Threats."** This title reflects the dual imperative for semiconductor companies to not only pursue customer diversification but also to critically assess and manage the risks posed by substitute technologies.

In practice, while the Total Addressable Market (TAM) projected by primary customers might suggest ample growth opportunities, it is essential to evaluate the long-term competitive viability of these customers. Semiconductor companies must conduct a comprehensive analysis that considers the competitiveness of the customer's own market segments (i.e., their secondary customers) and **the potential for substitute technologies to erode the primary customer's market position**. Overestimating the market potential of a customer based solely on initial demand projections can lead to unexpected challenges. Therefore, a multifaceted analysis using the extended 5 forces framework is indispensable.

Through this model, semiconductor companies are encouraged to ask critical questions: Have you encountered situations where over reliance on a customer's TAM and demand led to unanticipated challenges? How do the competitive dynamics and substitute threats compare across different customer segments? Such reflections enable companies to derive valuable insights and adjust their diversification strategies accordingly.

Model 9.
Diversified FAB Strategy for Risk Mitigation

1) Business Model analysis through The Extended 5 Forces

In the semiconductor industry, maintaining a stable supply chain is critical. Occasionally, internal circumstances necessitate a change in the production site, commonly referred to as the FAB (Fabrication) site. **Since the FAB site plays a pivotal role in defining the technical and quality characteristics of semiconductor products, any change demands meticulous preparation and rigorous verification**. This model explores situations in which FAB diversification is pursued not because of customer requirements, but rather due to internal strategic imperatives aimed at enhancing supply stability and process performance.

Typically, a change in the FAB site is implemented to secure supply continuity or to expand production capacity—especially in cases where the existing FAB undergoes process changes that reduce output, or when escalating customer demand strains the current facility. For new product introductions, companies may opt to initiate production at a new FAB rather than relying solely on the historical site, either to secure additional capacity or to adopt an improved manufacturing process that boosts product performance.

However, when the decision to change a FAB is driven by the semiconductor company's internal factor rather than by direct customer demand, significant challenges may arise. In such cases, it becomes imperative to carefully select the customers, models, and products for which the FAB change will be implemented, in order to minimize operational risks and facilitate effective customer persuasion.

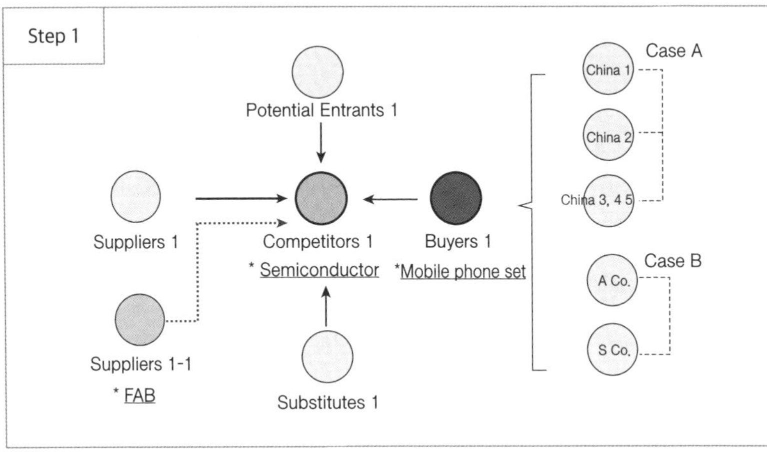

Figure 14 : Extended 5 forces framework for model 9

Figure 14 displays the extended 5 forces framework as applied to Model 9. In this analysis, the focus is on Competitors 1 (the semiconductor companies) that previously sourced products from Suppliers 1's FAB. In order to secure supply stability or enhance process performance, these companies diversified their FAB sources by incorporating a new FAB—denoted as Suppliers 1-1. At this juncture, the company must decide **which customers among Buyers 1 to target and which products to manufacture at the new FAB**, all while persuading these customers about the benefits of the transition.

Consider two hypothetical cases,

- ✓ **Case A**: Here, the FAB site change affects five mobile phone set customers (labeled China 1 through 5). Although having multiple customers might seem to hedge against risk, the concentration of these customers in **a single geographic region (China)** means that their demand fluctuations are highly correlated. Consequently, a simultaneous increase or decrease in demand across all these customers can lead to a significant mismatch—either a shortage or an oversupply relative to the company's production capacity.
- ✓ **Case B**: In this scenario, there is a leading mobile phone set company **(A Co.)** operating in the global flagship market, and another set company **(S Co.)** demonstrating strong competitiveness in the volume segment of the global smartphone market. Because **these customers operate on a global scale** rather than being regionally concentrated, the demand risk is more effectively dispersed, reducing the likelihood of extreme volatility.

The analysis of these two cases underscores the importance of comprehensively assessing the geographic distribution, strategic objectives, and demand volatility of the set customers when formulating a FAB diversification strategy. In addition, it is vital to include a broad range of products in the customer's portfolio—

from flagship models with stringent technical requirements to high-volume, mainstream models that offer stable demand. Rather than concentrating on a single product line, **semiconductor companies should develop a diversified FAB portfolio that enhances resilience against fluctuations in demand**.

When planning a large-scale FAB change at the node level, it is advisable to target well-established customers and projects. Thorough persuasion and mutual understanding between the semiconductor company and its customers are critical, as overly optimistic expectations regarding product quality and demand can precipitate significant failures.

2) Key Force and Market Diversification Strategies

Table 16 : Extended 5 forces and market diversification factors for model 9

Component	Factor		Business Model
Extended 5 Forces Theory	Key Force	Bargaining power of suppliers	◎
	Strength of the Competitive Forces for Semiconductor		◎
Competitors 1 analysis (Semiconductors)	**Technological innovation**	Technological innovation	○
		Successful transition from research to development	○
		Enhancing yield	○
	Diversification	Customer	△
		Application	△
		Product	◎

Table 16 summarizes the extended 5 forces and market diversification factors for Model 9. The analysis indicates that the key competitive element in this context is the bargaining power of suppliers. This is because supply shortages or the need for process improvements in the FAB can critically influence a semiconductor company's ability to meet market demand. Consequently, the decision to change a production site can be viewed as a form of product diversification aimed at mitigating risk.

This business model is termed the **"Diversified FAB Strategy for Risk Mitigation".** The strategy extends beyond merely changing production sites or expanding capacity; it emphasizes minimizing risk by enabling a flexible, multi-source supply chain. By diversifying its FAB sources, a semiconductor company can quickly adapt to rapidly evolving customer demands and technological advancements, thereby safeguarding its competitiveness against potential supply interruptions or quality issues that might arise from over-reliance on a single FAB.

Model 10.
EPI Map (Evaluation-Performance Insights Map)

1) Business Model Analysis through The Extended 5 Forces

In the B2B semiconductor industry, periodic **customer evaluations** serve as an essential mechanism for assessing and improving business performance. These evaluations provide critical insights into both strengths and weaknesses in the relationship between semiconductor companies and their customers, ultimately guiding strategies to enhance customer satisfaction and long-term profitability.

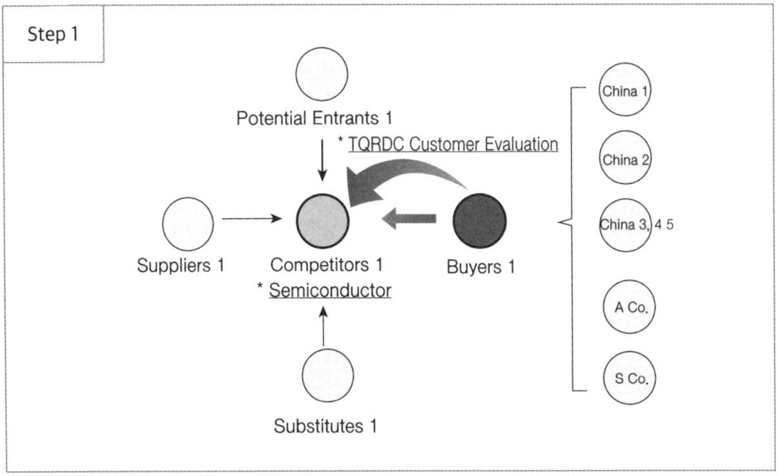

Figure 15 : Extended 5 forces framework for model 10

In this context, the extended 5 forces framework is applied primarily to analyze the evaluation practices performed by Buyers 1 (direct customers). Because the core focus here is on these evaluations, further subdivision into additional layers of the extended 5 forces is not applied, as illustrated in **Figure 15.**

Among the various methods available for conducting customer evaluations, I propose using the **TQRDC framework**—which stands for Technology, Quality, Responsiveness, Delivery, and Cost—as a foundational metric. This framework provides a multi-dimensional approach to evaluating the performance of semiconductor suppliers and offers a structured way to assess the quality of their business relationships.

The importance of TQRDC evaluation in the semiconductor industry is multifaceted. Semiconductor companies operate in an environment where different functions (such as development, purchasing, and quality assurance) interact closely. For instance, development teams collaborate with their counterparts at the customer's site, purchasing teams coordinate with sales teams, and quality control units work with corresponding quality assurance departments. Periodic, comprehensive evaluations conducted by customers enable both parties to review their interactions, identify deficiencies, and pinpoint areas for improvement. Such systematic evaluations are essential for refining semiconductor business models.

Specifically, TQRDC serves several critical roles,

✓ **Technology and Quality:** These dimensions allow customers to assess how effectively the semiconductor company meets their demands for technological innovation and product excellence. In rapidly evolving markets, a high score in these areas is a strong indicator of a company's technological capability and its commitment to continuous improvement. Companies that consistently deliver innovative solutions are likely to secure higher ratings, thereby unlocking additional business opportunities.

✓ **Cost:** In an industry where production efficiency is paramount, the cost component is crucial. Customers use this metric to determine whether a supplier can deliver high-quality products while maintaining cost efficiency (which, from the supplier's perspective, corresponds to **the price proposed to the customer**). Effective cost management not only boosts profitability for the supplier but also fosters long-term partnerships with customers who are under constant pressure to reduce costs.

✓ **Delivery:** Given the complexity and lengthy production cycles characteristic of semiconductor manufacturing, reliable delivery performance is indispensable. Evaluating delivery metrics provides insights into both short-term responsiveness and long-term operational stability. Consistent, timely delivery builds customer confidence and solidifies the supplier's reputation in the market.

- ✓ **Responsiveness:** This aspect measures a supplier's ability to adapt to changing requirements and resolve issues promptly. In a fast-paced and highly competitive industry, responsiveness is critical for maintaining customer satisfaction and sustaining business growth.

To leverage customer evaluation results effectively, semiconductor companies must integrate TQRDC outcomes with financial performance indicators. Although it might be intuitively expected that higher TQRDC scores would correspond with superior financial performance, empirical observations have shown that **this relationship is not always linear**. For example, some customers with high financial performance may exhibit lower TQRDC scores, potentially due to increased dependency on the supplier—which in turn might lead to higher dissatisfaction if expectations are not met. This phenomenon reflects the high bargaining power of suppliers within the 5 forces framework.

Recognizing this complexity, I propose the development of an integrated analytical tool: the **EPI Map (Evaluation-Performance Insights Map)**. This framework systematically combines **customer evaluations (via TQRDC), financial performance** metrics, and broader **customer analysis**. By mapping customer TQRDC scores against financial performance (such as profit and market share), the EPI Map categorizes customers into distinct segments. This graphical tool facilitates the identification of tailored strategies for each customer segment and enables the company to fine-tune its relationship management and pricing strategies.

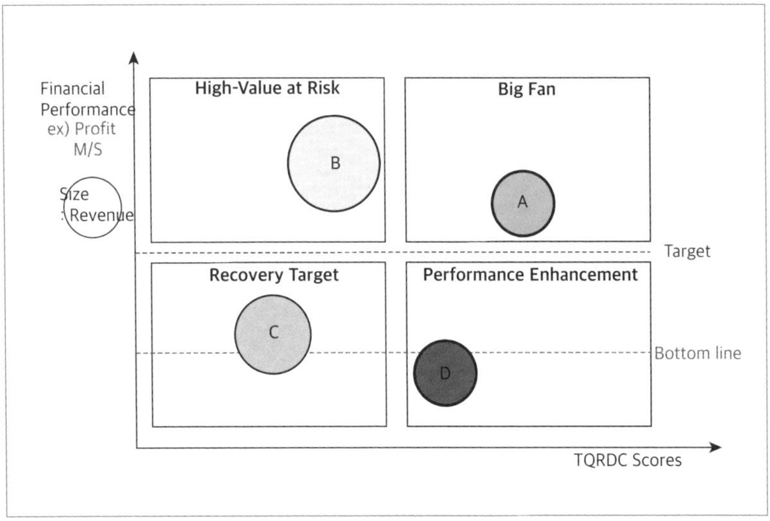

Figure 16 : EPI map and customer analysis

Figure 16 provides a visual representation of the EPI Map. In this graph, the Y-axis represents key financial performance metrics (e.g., operating profit or market share), while the X-axis captures the TQRDC evaluation scores assigned by customers. Each customer is represented by a bubble, whose size corresponds to its revenue contribution. The segmentation criteria on the X-axis are adjustable, allowing companies to define thresholds based on their strategic priorities. This mapping process yields several distinct customer categories, each with unique strategic implications.

The EPI Map typically identifies **four customer types**,

a) **Big Fan (Customer A)**

Characterized by very high TQRDC scores and significant profit contributions, these customers are considered ideal and loyal partners. Their satisfaction with product quality, technological innovation, and service consistency is a critical asset. Maintaining and further strengthening these relationships is vital for ensuring long-term stability and sustained revenue growth.

b) **High-Value at Risk (Customer B)**

Although these customers generate substantial revenue, their TQRDC scores are relatively low, signaling potential dissatisfaction or unresolved issues. Despite their financial importance, these customers represent a risk to long-term profitability if their concerns are not addressed through targeted strategic initiatives.

c) **Recovery Target (Customer C)**

These customers currently contribute modest profits and have low evaluation scores, indicating a weak relationship. However, they also possess considerable revenue potential.

If the underlying issues are resolved and the relationship is improved, these customers could become significantly more profitable. Strategic efforts focused on relationship recovery are essential in this category.

d) **Performance Enhancement (Customer D)**

This segment comprises customers with high TQRDC scores but relatively low financial contributions. Their satisfaction indicates that the supplier is delivering quality and innovation; however, issues such as pricing or insufficient market penetration may be limiting their profitability. In such cases, a reassessment of the business strategy is warranted to boost financial performance.

The EPI Map serves as a vital tool for semiconductor companies by providing comprehensive insights that integrate customer evaluations with financial performance data. This integrated approach enables the development of tailored strategies that enhance customer relationships, improve operational performance, and ultimately drive superior financial outcomes.

2) Key Force and Market Diversification Strategies

Table 17 : Extended 5 forces and market diversification factors for model 10

Component	Factor		Business Model
Extended 5 Forces Theory	Key Force	Bargaining power of buyers	◎
	Strength of the Competitive Forces for Semiconductor		○
Competitors 1 analysis (Semiconductors)	**Technological innovation**	Technological innovation	○
		Successful transition from research to development	○
		Enhancing yield	○
	Diversification	Customer	◎
		Application	○
		Product	○

Table 17 summarizes the extended 5 forces and market diversification factors for Model 10. In this analysis, the key competitive element identified is the bargaining power of buyers, which actively evaluates semiconductor companies through frameworks such as TQRDC. This model is formally designated as the **"EPI Map (Evaluation-Performance Insights Map)"** strategy.

By leveraging TQRDC evaluation results alongside financial performance metrics, semiconductor companies are empowered to reexamine their customer relationships and implement continuous improvements. This integration facilitates the development of bespoke strategies tailored to the unique characteristics of each customer segment. As a result, companies can foster stable and reliable long-term partnerships while simultaneously enhancing overall financial performance.

The EPI Map framework not only supports strategic customer management but also serves as a dynamic tool for enhancing financial performance and adapting to shifts in customer expectations over time. By mapping the interplay between customer satisfaction and financial outcomes, semiconductor companies can better anticipate challenges and seize emerging opportunities.

Chapter 4

Conclusion

Chapter 4

Conclusion

This book offers a comprehensive set of **strategic insights** for achieving **sustainable success** in the semiconductor industry. At its core, it emphasizes the **distinctive business models and key principles** that animate this technologically driven yet intensely competitive sector. Drawing on extensive professional experience and a close examination of emerging trends, I have endeavored to illuminate how traditional concepts, such as Porter's competitive strategy, require meaningful adaptation to address the semiconductor industry's intricacies. By illustrating these adaptations and proposing a series of frameworks—most notably **the extended 5 forces theory** and **market diversification** approach—this text aspires to guide current and future practitioners in **bridging the divide** between technology breakthroughs and business profitability.

A Multi-Layered Understanding of Competition

To thoroughly dissect the competitive structure of semiconductors, this book introduced the extended 5 forces theory, expanding Porter's original model to incorporate secondary customers alongside primary ones. This advancement allows for multi-tiered insight into how technological developments, supply chain interdependencies, and specialized functional areas converge to shape the market environment. **By spotlighting the customers' customers**, companies can detect indirect pressures or substitute threats that might otherwise remain hidden. In essence, adopting **the extended 5 forces perspective** enables semiconductor enter-prises to anticipate not just direct rival actions but also underlying shifts in downstream applications, usage patterns, or consumer preferences in semiconductor ecosystem dynamics. Such a broadened viewpoint is indispensable in an industry characterized by rapid iterative enhancements and intricate global supply chains.

Concurrently, the **competition** in semiconductors was shown to be evolving toward **market diversification.** Given the sector's turbulent demands—caused by cyclical consumer electronics, explosive new applications like AI, and variable regional growth patterns—companies must actively develop strategies that diversify their customer portfolios, product lines, and application domains. By mitigating reliance on a narrow set of markets, semiconductor firms stand better poised to exploit emerging opportunities and reduce vulnerabilities when key client groups or product lines approach stagnation. This holistic approach to diversification underscores a central theme in the book: the most resilient semiconductor companies do not passively wait for technology to dictate their fate; they explore multiple paths to growth, forging new alliances and cultivating fresh revenue streams in parallel with established competencies.

Recognizing the Power of Invisible Forces

Another significant thread woven through the book is the concept of **invisible forces**—factors not overtly included in standard strategic frameworks, yet powerful enough to sway entire market trajectories. The rise of **AI**—highlighted as a transformative example—demonstrates how an initially speculative technology can abruptly reshape value chains, upend conventional technology road maps, and intensify competitive pressures. **By framing these invisible forces as elements that orbit just beyond the extended 5 forces,** semiconductor companies are encouraged to remain vigilant and adaptive, scanning not only for immediate competitor moves but also for **emergent catalysts capable of reshaping markets.**

Keeping pace with such uncertain influences demands an openness to constant learning and iterative strategy refinement. A semiconductor player might anticipate a new wave of demand for specialized AI chips or innovative sensor modules, but if it fails to invest in the right design capabilities or partnerships early, it risks losing ground to agile rivals. Hence, **invisible forces do more than occasionally disrupt the status quo—they often define the next competitive frontier**.

Leveraging Porter's Strategy and Industry Experience

In my professional journey throughout the semiconductor business, I have consistently relied on Michael Porter's competitive strategy for guiding core decisions, from forging customer relationships to orchestrating complex partnerships. This theoretical baseline has proven instrumental in clarifying how to engage with formidable buyers, partner with specialized suppliers, and navigate incessant technological leaps. Yet my experiences also repeatedly illustrated that semiconductors require deeper layers of analysis than Porter's traditional model alone could provide. The intricate supply chains, rapid product life cycles, and specialized technology nodes central to semiconductors call for expanded theoretical perspectives —most notably, the Extended 5 Forces Theory.

Hence, the **10 business models** and **10 key principles** enumerated throughout this book primarily reflect the **deeper insight** I have experienced through direct and indirect involvement and analysis of real-life successes and failures in **semiconductor ecosystem dynamics**.

For instance, diversification strategies relating to customers and applications derive from the fundamental awareness that no single market or client can remain static forever. Similarly, the emphasis on yield, manufacturing proficiency, and strategic coordination with ecosystem partners underscores how intangible operational details can become decisive competitive advantages.

Sustainable Success in Semiconductor Ecosystem Dynamics

Drawing from my experiences and past challenges, I propose that the notions of insight and patience serve as fundamental pillars in achieving sustainable success.

a) **Insights** themselves are rarely generated by spotlighting one facet in isolation. They emerge from evaluating multiple angles, questioning assumptions, and contextualizing patterns across engineering, supply chain operations, and business strategy. A new chip design may be technically advanced, but if a company overlooks shifts in the end-customer's competitive environment or fails to respond to a new wave of AI-based demand, the product's advantage can be undercut. **True insight mandates an integrated, multi-perspective mindset** that extends beyond immediate R&D feats or near-term profit margins.

b) **Patience,** in turn, spotlights the requirement of sustained effort, unwavering dedication, and time. The semiconductor industry often demands more than a single product cycle or incremental process node improvement to achieve transformative gains. Many breakthroughs—and, indeed, many major successes—entail at least a decade of methodical refinement, forging alliances, stabilizing yields, and systematically honing a technology's cost structure. Thus, "Patience" denotes not simply perseverance but also the **vision** to endure and continuously adapt in the face of adversity. **"Time coupled with patience is an asset in semiconductor ecosystem dynamics."**

"Sustainable Success" further underscores the **aspiration** to overcome the inherent complexities, short-term setbacks, and cyclical downturns of this challenging sector. Ultimately, **harnessing robust business models,** guided by the **frameworks laid out in this book,** becomes the keystone in fostering truly lasting success.

Integrating Business Models, Principles, and Future Trajectories

Through each of the 10 models, I highlight distinct factors that semiconductor firms frequently encounter: extended supply chains, commoditizing technologies, specialized customers, invisible market shifts, or new forms of cross-partnership synergy. **The impetus behind these models is not to claim universal formulas but to showcase how companies can systematically evaluate and respond to repeated patterns.** In such a dynamic environment, even partial knowledge gleaned from a previous success or failure can illuminate the path for new technologies or customers. Indeed, new scenarios often harbor parallels with older experiences, albeit disguised by updated product specs or re-branded end markets.

Looking Ahead: A Personal Pledge

As a professor and active contributor to semiconductor strategy, I remain committed to refining the theories and methodologies advanced in this book. Technology evolves in leaps, and business landscapes may shift with equal rapidity—driven by consumer tastes, emergent ecosystems, or breakthroughs in computing paradigms. My aim, then, is to keep pace with these changes and update the foundational strategies accordingly. Whether a new dimension arises from emergent quantum computing applications, advanced packaging techniques, or future breakthroughs akin to AI, I plan to scrutinize how these developments:

a) **Conform** to the existing business model theories introduced here,

b) **Extend** or evolve from previously described structures, or

c) **Manifest** in previously unforeseen ways—demanding fresh analytical tools or expansions to the extended 5 forces or invisible forces concepts.

By systematically investigating these novel directions, we collectively reinforce the semiconductor ecosystem's ability not only to adapt to new demands but also to proactively shape them. Companies that assimilate the frameworks and principles explored in these pages will be better prepared to anticipate shifts, innovate robustly, and maintain competitiveness, even amid upheavals.

A Call to the Ecosystem

My conviction is that **genuine, sustainable success** in semiconductors stems from **orchestrating top-tier technology development** with **business acumen, supply chain mastery,** and **adaptive collaboration.** This entails forging stronger partnerships—both within one's internal teams (engineering, marketing, and executive leadership) and throughout the broader ecosystem (fabs, module integrators, IP providers, set manufacturers, potential new markets, and more). By unifying these facets, we can propel industry growth while also serving the growing needs of global customers. In other words, **success** is not measured simply in product shipments or short-term profitability but also **in stable, trust-based networks that thrive over the long term, driven by semiconductor ecosystem**

dynamics. In this context, **"networks"** denote strategically oriented, **trust-based relationships across the semiconductor ecosystem—built on mutual understanding, sustained collaboration, and long-term value creation.**

With unwavering commitment, I pledge to enrich and enhance my own research into semiconductor business strategies. By advocating for these frameworks in academic circles, fostering active dialogue with colleagues in the field, and continuously refining my approach in light of evolving market realities, I aspire to see the entire semiconductor ecosystem flourish. When truly strategic directions are established, participants can lead, rather than merely respond to, the next wave of changes and challenges in semiconductor ecosystem dynamics.

Concluding Reflections

In closing, let us re-emphasize the central question driving this book:

"How do semiconductor companies succeed?"

The solutions do not lie in single-factor analysis or short-term maneuvers; they instead require **a multi-level perspective** and **insights into business models grounded in established patterns and principles**. Furthermore, it is imperative to craft **a competitive strategy** that fosters synergy among technology, business models, and market forces. By incorporating models like **the extended 5 forces**, recognizing the inevitability of **market diversification,** spotting **invisible forces,** and effectively leveraging **the 10 business models** in practice, companies can elevate their strategies beyond incremental competition and move toward sustainable leadership.

Ultimately, **the great journey toward sustainable success** is marked by **perseverance** in face of adversity, **trust-based networks**, and **relentless openness** to learning from both triumphs and failures. And while patience is vital, equally important is the ability—to convert that patience into targeted, proactive initiatives that secure real, lasting achievements. May these insights serve as a steadfast foundation for all members of the semiconductor ecosystem—engineers, managers, executives, students, and researchers—who strive to make thoughtful decisions that align cutting-edge technology with viable, meaningful, and profitable business outcomes.

References

1. Porter, M. E. (1980). Competitive strategy: Techniques for analyzing industries and competitors. New York, NY: Free Press.
2. Porter, M. E. (1985). Competitive advantage: Creating and sustaining superior performance. New York, NY: Free Press.
3. Porter, M. E. (2008). On competition. Boston, MA: Harvard Business Review Press.
4. Winston, P. H. (1984). Artificial intelligence. Reading, MA: Addison-Wesley.
5. Winston, P. H. (2019). The AI business: The commercial uses of artificial intelligence. Cambridge, MA: MIT Press.
6. Abernathy, W. J., & Utterback, J. M. (1978). Patterns of industrial innovation. Technology Review, 80(7), 40-47.
7. Levitt, T. (1965). Exploit the product life cycle. Harvard Business Review, 43(6), 81-94.
8. Kotler, P. (1997). Marketing management: Analysis, planning, implementation, and control. Upper Saddle River, NJ: Prentice Hall.
9. Christensen, C. M. (1997). The innovator's dilemma: When new technologies cause great firms to fail. Boston, MA: Harvard Business Review Press.

10. Christensen, C. M., & Raynor, M. E. (2003). The innovator's solution: Creating and sustaining successful growth. Boston, MA: Harvard Business Review Press.
11. Ansoff, H. I. (1957). Strategies for diversification. Harvard Business Review, 35(5), 113-124.
12. Teece, D. J. (2007). Explicating dynamic capabilities: The nature and microfoundations of (sustainable) enterprise performance. Strategic Management Journal, 28(13), 1319-1350.
13. Dyer, J. H., & Singh, H. (1998). The relational view: Cooperative strategy and sources of interorganizational competitive advantage. Academy of Management Review, 23(4), 660-679.
14. Grant, R. M. (1996). Toward a knowledge-based theory of the firm. Strategic Management Journal, 17(S2), 109-122.
15. Rumelt, R. P. (1991). How much does industry matter? Strategic Management Journal, 12(3), 167-185.